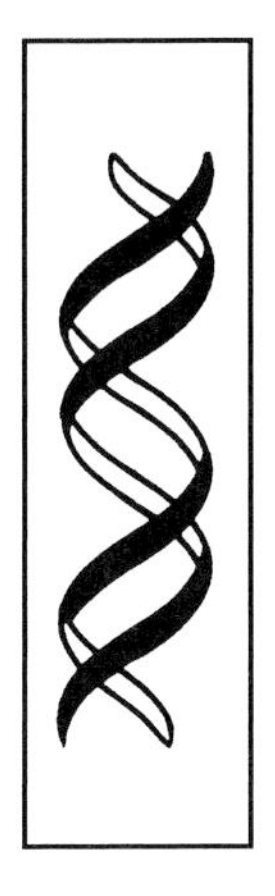

Computational Methods for Genetics of Complex Traits

Advances in Genetics, Volume 72

Serial Editors

Theodore Friedmann
University of California at San Diego, School of Medicine, USA

Jay C. Dunlap
Dartmouth Medical School, Hanover, NH, USA

Stephen F. Goodwin
University of Oxford, Oxford, UK

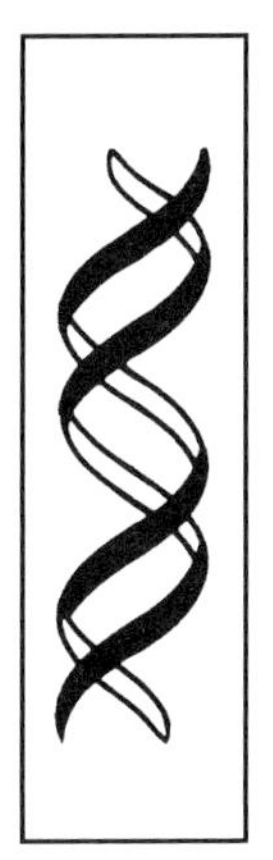

Volume 72

Computational Methods for Genetics of Complex Traits

Edited by

Jay C. Dunlap

Department of Genetics
Dartmouth Medical School
Hanover, NH, USA

Jason H. Moore

Institute for Quantitative Biomedical Sciences
Departments of Genetics and Community and Family Medicine
Dartmouth Medical School
Lebanon, NH, USA

ELSEVIER

AMSTERDAM • BOSTON • HEIDELBERG • LONDON
NEW YORK • OXFORD • PARIS • SAN DIEGO
SAN FRANCISCO • SINGAPORE • SYDNEY • TOKYO
Academic Press is an imprint of Elsevier

Academic Press is an imprint of Elsevier

525 B Street, Suite 1900, San Diego, CA 92101-4495, USA
30 Corporate Drive, Suite 400, Burlington, MA 01803, USA
32 Jamestown Road, London, NW1 7BY, UK
Radarweg 29, POBox 211, 1000 AE Amsterdam, The Netherlands

First edition 2010

ISBN: 978-0-12-380862-2
ISSN: 0065-2660

For information on all Academic Press publications
visit our website at elsevierdirect.com

Printed and bound in USA

10 11 12 10 9 8 7 6 5 4 3 2 1

Contents

Contributors

Numbers in parentheses indicate the pages on which the authors' contributions begin.

James W. Baurley (47) Department of Preventive Medicine, University of Southern California, Los Angeles, California, USA

William S. Bush (1) Center for Human Genetics Research, Vanderbilt University, Nashville, Tennessee, USA, and Department of Biomedical Informatics, Vanderbilt University, Nashville, Tennessee, USA

David V. Conti (47) Department of Preventive Medicine, University of Southern California, Los Angeles, California, USA

Robert Culverhouse (117) Washington University in St Louis School of Medicine, St Louis, Missouri, USA

Peter Holmans (141) Biostatistics and Bioinformatics Unit, MRC Centre for Neuropsychiatric Genetics and Genomics, Department of Psychological Medicine and Neurology, Cardiff University School of Medicine, Heath Park, Cardiff, United Kingdom

Sharon L. R. Kardia (181) Department of Epidemiology, School of Public Health, University of Michigan, Ann Arbor, Michigan, USA

Reagan J. Kelly (181) Division of Bioinformatics, ICF International at National Center for Toxicological Research, U.S. Food and Drug Administration, Jefferson, Arkansas, USA

Jason H. Moore (101) Institute for Quantitative Biomedical Sciences, Departments of Genetics and Community and Family Medicine, Dartmouth Medical School, Lebanon, New Hampshire, USA

Marylyn D. Ritchie (1) Center for Human Genetics Research, Vanderbilt University, Nashville, Tennessee, USA, and Department of Molecular Physiology and Biophysics, Vanderbilt University, Nashville, Tennessee, USA; and Department of Biomedical Informatics, Vanderbilt University, Nashville, Tennessee, USA

Ingo Ruczinski (25) Department of Biostatistics, Bloomberg School of Public Health, Johns Hopkins University, Baltimore, Maryland, USA

Holger Schwender (25) Department of Biostatistics, Bloomberg School of Public Health, Johns Hopkins University, Baltimore, Maryland, USA

Jennifer A. Smith (181) Department of Epidemiology, School of Public Health, University of Michigan, Ann Arbor, Michigan, USA

Yan V. Sun (73) Department of Epidemiology, School of Public Health, University of Michigan, Ann Arbor, Michigan, USA

Duncan C. Thomas (47) Department of Preventive Medicine, University of Southern California, Los Angeles, California, USA

Melanie A. Wilson (47) Department of Preventive Medicine, University of Southern California, Los Angeles, California, USA

1

Genome Simulation: Approaches for Synthesizing *In Silico* Datasets for Human Genomics

Marylyn D. Ritchie*,†,‡ and William S. Bush*,‡

*Center for Human Genetics Research, Vanderbilt University, Nashville, Tennessee, USA

†Department of Molecular Physiology and Biophysics, Vanderbilt University, Nashville, Tennessee, USA

‡Department of Biomedical Informatics Vanderbilt University, Nashville, Tennessee, USA

Advances in Genetics, Vol. 72

0065-2660/10 $35.00
DOI: 10.1016/S0065-2660(10)72001-0

ABSTRACT

Simulated data is a necessary first step in the evaluation of new analytic methods because in simulated data the true effects are known. To successfully develop novel statistical and computational methods for genetic analysis, it is vital to simulate datasets consisting of single nucleotide polymorphisms (SNPs) spread throughout the genome at a density similar to that observed by new high-throughput molecular genomics studies. In addition, the simulation of environmental data and effects will be essential to properly formulate risk models for complex disorders. Data simulations are often criticized because they are much less noisy than natural biological data, as it is nearly impossible to simulate the multitude of possible sources of natural and experimental variability. However, simulating data *in silico* is the most straightforward way to test the true potential of new methods during development. Thus, advances that increase the complexity of data simulations will permit investigators to better assess new analytical methods. In this work, we will briefly describe some of the current approaches for the simulation of human genomics data describing the advantages and disadvantages of the various approaches. We will also include details on software packages available for data simulation. Finally, we will expand upon one particular approach for the creation of complex, human genomic datasets that uses a forward-time population simulation algorithm: genomeSIMLA. Many of the hallmark features of biological datasets can be synthesized *in silico*; still much research is needed to enhance our capabilities to create datasets that capture the natural complexity of biological datasets.

I. INTRODUCTION

The ultimate goal in understanding the biology of human disease is to identify all the genomic variations relevant to the phenotype or trait being studied. As molecular genomic technology has advanced, the field has gone from very coarse genomic examination embodied in cytogenetic analyses, to higher resolution linkage analyses, and now to very high resolution association analyses with up to billions of data points per experiment. The ultimate resolution of having complete individual genome sequences is on the horizon as a stated goal of the National Human Genome Research Institute. Methodological advances in the analysis of large-scale genome-wide association studies (GWAS), including the ability to integrate results across multiple experiments, have simply not kept pace with advances in data acquisition. In fact, most GWAS studies have taken a simplistic approach to analysis, considering the effect of each single SNP on the phenotype. Many susceptibility genes may exhibit effects that are partially or solely dependent on interactions with other genes and/or the environment, and while these complex effects are beginning to be explored in GWAS data, methods to detect and model the complexity presumed to be a part of common disorders are still in their infancy.

To deal with the challenge of detecting complex genetic effects, much research is underway to develop improved statistical and computational methodologies. Many researchers are exploring variations and modifications of logistic regression such as logic regression (Kooperberg *et al.*, 2001), penalized logistic regression (Zhu and Hastie, 2004), classification/regression trees (CART), and multivariate adaptive regression splines (MARS) (Cook *et al.*, 2004). Additional studies are being conducted in data mining and machine learning research, including data reduction and pattern recognition approaches. Data reduction involves a collapsing or mapping of the data to a lower-dimensional space. Examples of data reduction approaches include the combinatorial partitioning method (CPM) (Nelson *et al.*, 2001), restricted partition method (RPM) (Culverhouse *et al.*, 2004), set association (Wille *et al.*, 2003), and multifactor dimensionality reduction (MDR) (Hahn *et al.*, 2003; Ritchie *et al.*, 2001; Ritchie *et al.*, 2003). Pattern recognition on the other hand, involves extracting patterns from the data to discriminate between groups using the full dimensionality of the data. Examples of pattern recognition methods include cluster analysis (Kaufman and Rousseeuw, 1990), cellular automata (CA) (Wolfram, 1994), support vector machines (SVM) (Cristianini and Shawe-Taylor, 2000), self-organizing maps (SOM) (Hastie *et al.*, 2001), and neural networks (NN) (Ripley, 1996). The application of these methods to human genetics data has recently been comprehensively reviewed (Motsinger *et al.*, 2007).

A fair comparison of these approaches requires a way to embed realistic probability models of human disease into large-scale genomic data in a manner that is reproducible and that allows for resampling to produce variability associated with natural and experimental processes. Many data simulation software packages have been developed with these goals in mind.

An alternative is to use experimental datasets in lieu of simulation; however, when natural biological data are used to test new methods and significant results are found, it is impossible to know if they are false positives or true positives. Similarly, if nothing significant is detected one cannot know if this is due to a lack of statistical power, or the data had no true signal to detect. Thus, it will be vital for the success of novel statistical and computational genomics methods to have the ability to simulate datasets consisting of polymorphisms throughout the genome on the scale of what is technically feasible. Having simulated data allows one to evaluate whether a methodology can detect known effects, and explore the range of genetic models and effect sizes that are detectable. Data simulations are often criticized because they are much less noisy than real data. However, simulating data remains an important component of most novel method-development projects. To this end, any advances to improve the complexity of the data simulations will permit investigators to better assess new analytical methods.

II. GENOME SIMULATION CHALLENGES

GWAS have quickly become the standard study design for disease gene discovery in both family-based and population-based samples. The GWAS design evolved from the work of the International HapMap Project, which aims to characterize common human variation across multiple ethnic subpopulations. The basic premise of GWAS, as well as candidate gene association studies, is that regions of the human genome are in linkage disequilibrium (LD) and single nucleotide polymorphisms (SNPs) or other genetic variations that lie in regions of high LD are inherited together on the same genomic segment through generations of human populations. If a genomic segment harbors a variation that incurs disease risk, other markers on the same genomic segment (in high LD) will show some degree of association to the disease indirectly (due to the correlation between the SNPs). As such, modern genotyping technologies attempt to either directly or indirectly (through LD) capture much of the common genetic variation in the human genome by typing hundreds of thousands to millions of SNPs scattered throughout the human genome.

Tools to simulate human genetic data sets are critical for the development of both basic and sophisticated GWAS data analysis approaches. There are three main challenges associated with simulating datasets with the features of underlying complexity expected in natural biological datasets: generating LD patterns similar to those in the human genome, generating LD patterns corresponding to different ancestral human populations, and generating realistic probability models for disease susceptibility.

First, it is known that patterns of LD are not uniform across the genome. As such, the degree of correlation among SNPs is dependent on its genomic location and cannot be mimicked using a single basic correlation structure. Suppose we model correlation among two SNPs as a function of the distance between them; this approach may work well in modeling one particular region of the genome, but it may or may not work as well for other areas of the genome. Additionally, LD patterns vary dramatically in different ancestral populations. Data from the International HapMap has documented that the LD patterns vary based on geographic region of ancestry, as well as degree of admixture in the population (Frazer *et al.*, 2007). As such, individuals with different geographic/ethnic ancestry have different LD structures, and in the case of admixed populations may have a blend of two distinct LD structures. This increases the difficulty in simulating a diverse genetic sample. Finally, because the primary type of association discovered thus far in the hundreds of GWAS published includes independent, single-locus effects, it is not entirely clear what types of genetic models should be simulated. As mentioned earlier, it is widely accepted that common diseases are inherently complex. Thus, we expect multiple genes as well as environmental factors to contribute to disease susceptibility. Unfortunately, because we do not have a wealth of successful complex disease models for common disease, it is not known what types of models will be most commonly observed in natural, biological datasets. Therefore, researchers performing data simulation experiments tend to create a variety of disease susceptibility models to survey the landscape of possible disease models.

III. GENERAL SIMULATION STRATEGIES

There are multiple strategies for simulating human genetic data. Approaches for the simulation of genetic data typically fall into one of four categories: probabilistic, coalescent, resampling, and forward-time. Each of these general strategies has benefits and weaknesses. We will explore each of these categories below followed by an expanded discussion of one of these categories: forward-time simulations.

A. Probabilistic simulations

In the late 1980s and early 1990s, as computer power became more accessible, several simulation packages were developed to assess the empirical power of family-based datasets for linkage analysis—the primary analysis and study design of the time (Table 1.1). These algorithms used probability models in conjunction with assumptions of Hardy–Weinberg equilibrium and Mendelian segregation expectations to create *in silico* families from which datasets can be drawn. Packages such as SIMLINK (Boehnke, 1986; Boehnke and Ploughman, 1997; Ploughman and Boehnke, 1989), SIMULATE (Boehnke, 1986; Boehnke and Ploughman, 1997; Ploughman and Boehnke, 1989), SLINK (Ott, 1989; Weeks *et al.*, 1990), and SIMLA (Bass *et al.*, 1993; Schmidt *et al.*, 2004a) are capable of simulating new pedigree–genotype structures from the existing genotypes of founder individuals in the original dataset. Many of these packages are also capable of simulating case–control data based on user-specified disease parameters. One of the advantages of probabilistic simulations is that the user maintains a great deal of control over the parameters and thus resulting datasets in the simulation. This control also leads to significant disadvantages to this approach. It can be very challenging to model complex scenarios as the parameters required to create such scenarios are unknown. Furthermore, the

Table 1.1. Simulation Packages

Type	Name	Reference
Probabilistic	SIMLINK	Boehnke (1986), Boehnke and Ploughman (1997), Ploughman and Boehnke (1989)
	SIMULATE/ SIMULATE2	Terwilliger *et al.* (1993),Terwilliger and Ott (1994)
	SLINK	Ott (1989), Weeks *et al.* (1990)
	SIMLA	Bass *et al.* (1993), Schmidt *et al.* (2004a)
Coalescent-based	GENOME	Liang *et al.* (2007)
	MS	Hudson (2002)
	SelSim	Spencer and Coop (2004)
	CoaSim	Mailund *et al.* (2005)
	FastCoal	Marjoram and Wall (2006)
	MaCS	Chen *et al.* (2009)
	MSMS	Ewing and Hermisson (2010)
Resampling-based	HAP-SAMPLE	Wright *et al.* (2007b)
	HAPSIMU	Zhang *et al.* (2008)
	GWAsimulator	Li and Li (2008)
	gs	Li and Chen (2008)
Forward-time-based	easyPOP	Balloux (2001)
	FPG	Hey (2005)
	FREGENE	Hoggart *et al.* (2007)
	simuPOP	Peng *et al.* (2007), Peng and Kimmel (2005)
	GenomePop	Carvajal-Rodriguez (2008)
	genomeSIMLA	Edwards *et al.* (2008)

parameters of the model may not fall within an intuitive framework for the user of the simulation package, for example, specifying penetrance values may be the preferred disease model representation for a classical geneticist, but specifying logistic model coefficients may be preferred for an epidemiologist. Furthermore, probability-based simulations tend to lack the same degree of noise and outliers seen in natural datasets, as values are drawn from a distribution with user-specified parameters. Additionally, these simulations tend to synthesize data of small to moderate scale in terms of number of SNPs, however, simplistic genetic models may be possible for larger scale datasets due to the computational simplicity of the methods.

B. Coalescent simulations

Coalescent models were developed in the mid 1990s to study population genetics phenomena such as haplotype divergence, mutation, and migration (Kingman, 1982). Coalescent models are retrospective, attempting to trace the flow of alleles backward through a population to a most recent common ancestor. Initially, coalescent-based simulations were utilized specifically by population geneticists for modeling evolutionary processes. In recent years, however, statistical geneticists have implemented the coalescent-model approach for simulating datasets for statistical genetics experiments. Several coalescent-based simulators have been implemented, such as MS (Hudson, 2002), SelSim (Spencer and Coop, 2004), CoaSim (Mailund *et al.*, 2005), and FastCoal (Marjoram and Wall, 2006). Due to the high computational load of coalescent-based simulation, this strategy is used primarily for short DNA sequences, but GENOME (Liang *et al.*, 2007) and MaCS (Chen *et al.*, 2009) use novel algorithms to simulate larger scale data with the coalescent. Additional advances have been made in coalescent simulations (MSMS) where the use of forward simulations can be incorporated with the coalescent simulation for selection dependent models (Ewing and Hermisson, 2010). The coalescent simulation has a significant strength in its ability to model features of population genetics extremely well, including mutation rates, genetic drift, and genealogical relationships. Coalescent simulations are also easily adaptable for the generation of sequence data, and can model complex population structures. A disadvantage of this approach is the computational load, as many of these software packages are hindered by processing power and memory requirements. Because of this, coalescent-based simulations were not typically used for simulating large-scale genomic datasets. Also, the retrospective design of the simulation complicates the selection of model parameters for the type of data desired.

C. Resampling simulations

In recent years, high-density natural, biological datasets have become publicly available resources. Because of this availability, several software packages have been developed to resample these existing data to synthesize additional datasets for data simulation experiments. HAP-SAMPLE uses phased chromosomes from HapMap data to generate new datasets (Wright *et al.*, 2007a), HAPSIMU uses sequenced ENCODE regions to generate heterogeneous population datasets (Zhang *et al.*, 2008), and GWAsimulator resamples existing genome-wide genotypes to generate data with similar LD patterns (Li and Li, 2008). Another software package, *gs*, quickly generates large datasets using LD patterns and haplotype blocks from the HapMap or any other natural, biological datasets available to the user (Li and Chen, 2008). These software packages have the advantage that the data clearly reflect LD patterns in the HapMap, since those data are used as a seed to initiate data simulation. This means that the datasets simulated using these approaches do possess the noisy nature of natural biological datasets, however, the variability of experimental noise will be limited. Similarly, because many of the datasets from which they sample (such as the HapMap populations) are limited in sample size and therefore number of unique chromosomes to sample from, there are clear limitations in the genetic diversity attainable from these simulations. Furthermore, using resampling approaches, LD patterns will always be modeled after populations with existing high-density data. As such, the effectiveness of methods on other populations with unknown LD patterns or admixture could not be assessed.

D. Forward-time simulations

As computer power and memory have expanded and their cost decreased, several forward-time simulation packages have been developed that attempt to model individuals in a population as they mate and progress forward through generations. Because they follow basic genetics theory, forward-time simulation approaches are easy to understand while retaining the flexibility to synthesize complex population effects and disease models for genomic datasets. Forward-time simulations marry the population genetics theory of the coalescent-model with the mathematical simplicity of probability models to produce novel, theoretically sound populations with distinct patterns of genetic variation and LD. The disadvantages of the forward-time approach are similar to other techniques—computational complexity and difficulty in specifying population and disease parameters.

Forward-time packages include easyPOP (Balloux, 2001), FPG (Hey, 2005), FREGENE (Hoggart *et al.*, 2007), simuPOP (Peng and Kimmel, 2005; Peng *et al.*, 2007), and GenomePop (Carvajal-Rodriguez, 2008). Much like other simulation approaches, many of these packages do not easily generate genome-wide genetic data; however, existing approaches are being modified, and new methods are being developed, such as genomeSIMLA.

IV. A FORWARD-TIME SIMULATION EXAMPLE: GENOMESIMLA

The motivation for developing genomeSIMLA was to achieve the ability to simulate: (1) realistic patterns of LD in human populations, (2) genome-wide association datasets in both family-based and case–control study designs, (3) single or multiple independent main effects, and (4) purely epistatic gene–gene interactions in efficient, user friendly software. Existing simulation packages can do one or more of these, but few are able to succeed in all areas. To achieve these goals, the forward-time simulation approach was used. In choosing this strategy, however, we did *not* aim to provide a realistic simulation of human evolution, though the software framework could potentially be used for that purpose.

GenomeSIM (Dudek *et al.*, 2006) was developed for the simulation of large-scale genomic data in population-based case–control samples. It is a forward-time population simulation algorithm that allows the user to specify many evolutionary parameters and control evolutionary processes. SIMLA (or SIMulation of Linkage and Association) (Bass *et al.*, 1993; Schmidt *et al.*, 2004b) is a simulation program that allows the user to specify varying levels of both linkage and LD among and between markers and disease loci.

SIMLA was specifically designed for the simultaneous study of linkage and association methods in extended pedigrees, but the penetrance specification algorithm can also be used to simulate samples of unrelated individuals (e.g., cases and controls). We have combined genomeSIM as a front-end to generate a population of founder chromosomes. This population will exhibit the desired patterns of LD that can be used as input for the SIMLA simulation of disease models; the combined algorithm and software package *genomeSIMLA*. Particular SNPs may be chosen to represent disease loci according to desired location, correlation with nearby SNPs, and allele frequency. Using the SIMLA method of disease modeling, up to six loci may be selected for main effects and all possible two- and three-way interactions as specified in (Marchini *et al.*, 2005) among these six loci are available to the user as elements of a disease model. Once these loci are chosen the user specifies disease prevalence, a mode of inheritance for each locus, and relative risks of exposure to the genotypes at each locus. An advantage of the SIMLA approach to the logistic function is it can simulate data on markers that are not independent, yet yield the correct relative risks and prevalence. Many simulation packages using a logistic function for penetrance specification do not have this capability. Modeling of purely epistatic interactions with no detectable main effects, as in genomeSIM, is also supported separately by an algorithm called SIMPEN (described in more detail below) and can simulate two-way, three-way, up to *n*-way interactions. Purely epistatic modeling allows the user to specify a model odds ratio, heritability, and prevalence for disease effects. Thus, the marriage of genomeSIM and SIMLA has allowed for the simulation of large-scale datasets with realistic patterns of LD and diverse realistic disease models in both family-based and case–control data.

A. Forward-time population simulation phase

GenomeSIMLA uses a forward-time population simulator which relies on random mating, genetic drift, recombination, and population growth to generate a population to with natural LD features (as shown in Fig. 1.1; Edwards *et al.*, 2008). Populations are haploid and chromosome-based—as such when individuals are created from the population, two chromosomes are sampled from to form a diploid individual. Simulated chromosomes are composed of biallelic SNP loci with user-specified positions in both physical and genetic distance, along

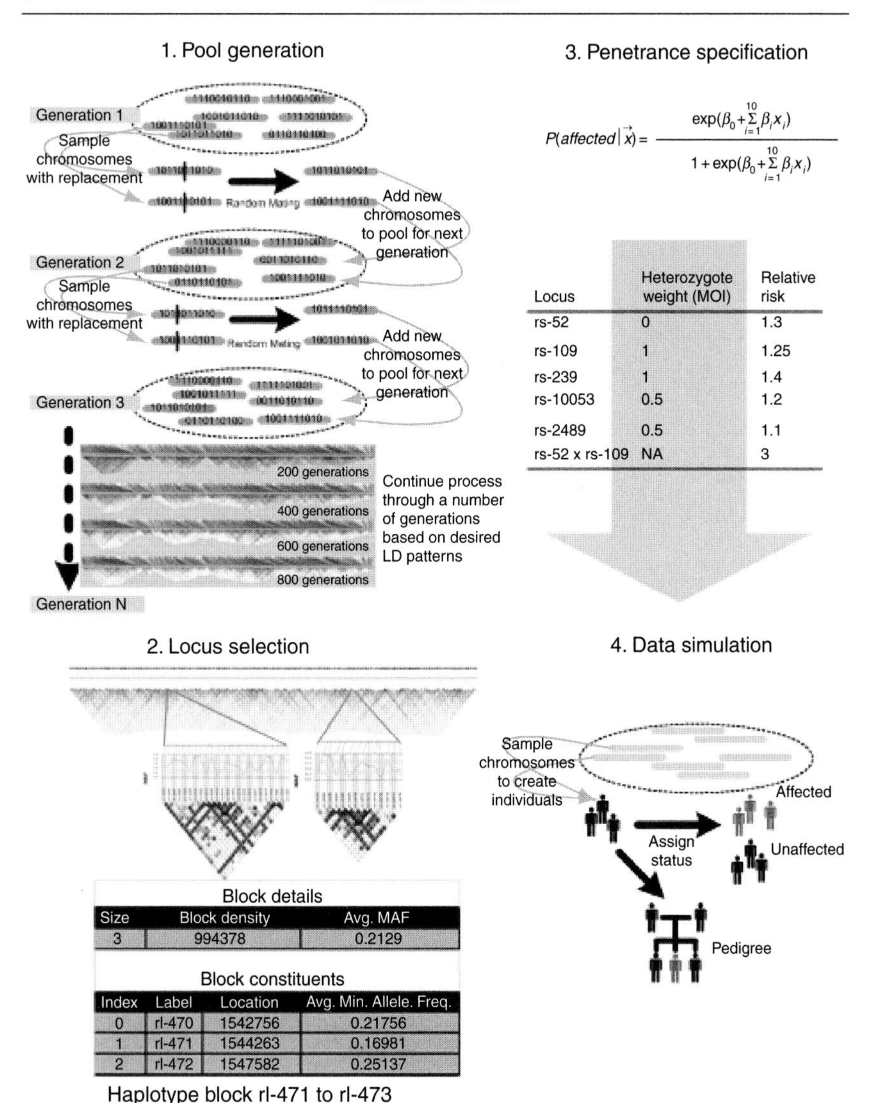

Figure 1.1. Overview of the simulation process.

with the allele frequency of each of these SNPs in the initial population. Positions and allele frequencies can easily be based on real SNPs using resources such as dbSNP or the HapMap project. The total number of SNPs per chromosome is unlimited except by hardware considerations, and multiple distinct groups of independent chromosomes can be created. Genome-wide association scale data can be readily generated by specifying 22 groups of chromosomes (one for each human autosome) and tens of thousands of SNPs per chromosome, such as the data collected from the Affymetrix 6.0 SNP Chip.

A pair of chromosomes is drawn from the population with replacement, and chromosomes are recombined. Chromosomal recombination occurs by drawing from a Poisson distribution with a lambda equal to the length in genetic distance (Morgans) of the chromosome. The recombination points on the chromosome are selected by randomly drawing positions (in genetic distance) from the uniform chromosome. Of the two resulting recombinant chromosomes, one is randomly chosen to advance to the next generation of the simulation. The random mating and recombination process continues on the population of chromosomes for a number of generations to generate realistic patterns of LD and produce sufficient numbers of chromosomes for drawing datasets. To achieve reasonable population sizes and to aid the formation of LD patterns, growth models are used to expand the population size dynamically over generations. Four growth models (three of which are shown in Fig. 1.2) are implemented to allow expansion of the population over generational time: linear, exponential, basic logistic, and the Richard's logistic curve. Each of these models also includes a variation parameter to add subtle fluctuation or jitter to the growth model.

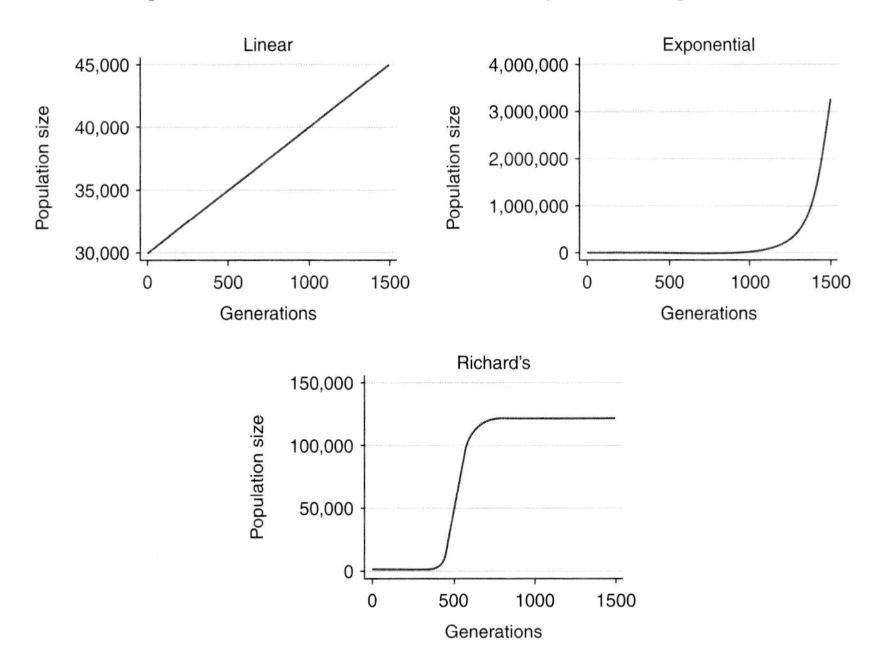

Figure 1.2. Available population growth functions.

Once the population has reached the user-specified size, or the population has advanced a user-specified number of generations, the chromosomal population is stored to disk in binary format, and the population simulation phase is complete.

B. Genetic effect simulation phase

After the pool of chromosomes has developed suitable LD and grown to a useful size, datasets can be drawn by randomly sampling chromosomes with replacement to create nonredundant individuals or families. Individuals or family members can then be assigned disease status based on simulated SNPs. Disease-susceptibility effects of multiple genetic variables can be modeled with user-specified penetrance functions. These functions can be entered manually, or generated using either the SIMLA logistic function (Schmidt *et al.*, 2004a) or a purely epistatic multilocus penetrance function found using simPEN, a genetic algorithm-based approach (Moore *et al.*, 2004).

SIMLA generates a penetrance function (or probability of disease given genotype) using a logistic regression equation:

$$\ln\left(\frac{p(\text{aff}|\vec{x})}{1 - p(\text{aff}|\vec{x})}\right) = \beta_0 + \sum_{i=1}^{N} \beta_i x_i \tag{1.1}$$

where x_i corresponds to a covariate value (either genetic or environmental) and β_i corresponds to the natural logarithm of the desired relative risks of disease due to a unit increase in x_i. Other user-specified parameters in the disease model include:

Parameter	Description
G_i	Allele frequency of allele at locus *i*
D_i	Penetrance, or probability of disease given allele G_i
W_i	Mode of inheritance at disease locus G_i where 1 is completely dominant, 0 is completely recessive, 0.5 is perfectly additive, and other values are mixtures of these models.
β_{jk}	The deviation in log relative risk from an additive combination of β_i and β_j per unit increase in interaction term $x_j x_k$, a multiplicative combination of genetic and/or covariate values *i* and *j*.
K	Disease prevalence in population

As seen above, the coding for the homozygous risk allele is always 1, and for the homozygous non-disease allele is always 0, and the coding for the heterozygote is somewhere between 0 and 1, as determined by the user-specified mode of inheritance parameters (W_i). This allows for a general group of inheritance models where the risk for the heterozygote is increased at some level between both homozygotes. Once all parameters have been specified, a slight modification of Newton's method is applied to find the value of β_0 that makes the model match the target prevalence.

SimPEN is an alternative strategy for assigning disease status to individuals. SimPEN was designed explicitly to generate multilocus penetrance models where there are no main effects from any single-locus alone; with such a genetic model, all loci must be evaluated jointly to detect the genetic effect (Moore *et al.*, 2004). In terms of heritability, these models have no additive or dominant variance—all of the trait variance due to genetics is explained by the nonadditive interaction of genotypes. Genetic models of this type were explored in (Culverhouse *et al.*, 2002), where these types of models are mathematically illustrated to potentially play a role in complex disease.

An outline of the simPEN procedure is shown in Fig. 1.3. SimPEN uses a genetic algorithm to generate a penetrance model meeting the specifications of the user. The model is arrived at by minimizing marginal penetrance variance to simulate a model with minimal main effects while also optimizing a function based on user-specified heritability, table variance (or overall variance of all penetrance values), and average marginal penetrance. This function is evaluated for each penetrance model in the population. Marginal penetrance variance and

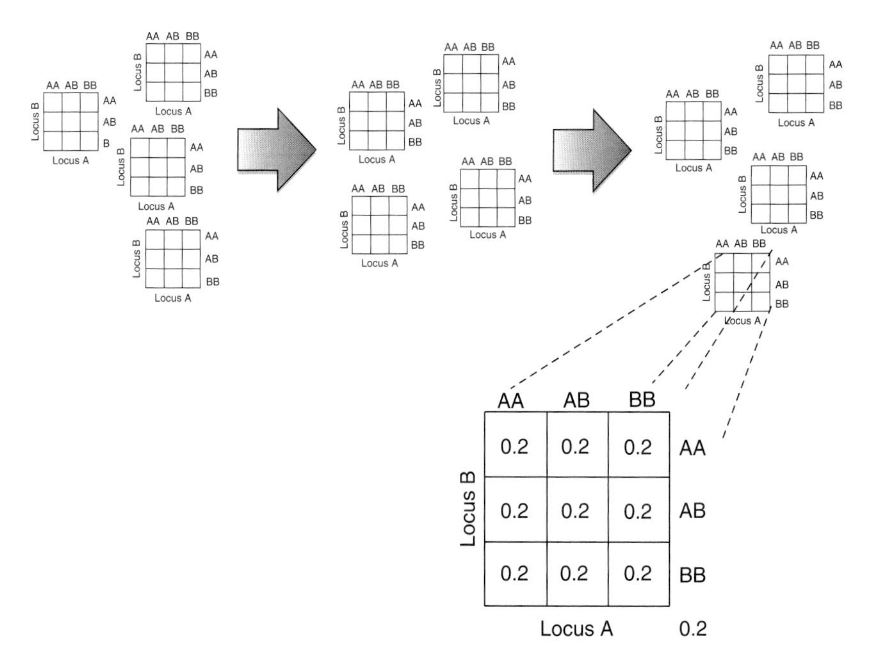

Figure 1.3. Overview of the simPEN procedure.

heritability always affect fitness, but users can optionally establish targets for table variance and marginal penetrance. The maximum fitness for this procedure is 1.0. The genetic algorithm terminates when it finds a model with fitness equal to 1.0 or the maximum number of generations is reached. Using this procedure, high-order multilocus genetic models exhibiting no main effect can be readily generated (Moore *et al.*, 2004).

C. Dataset generation phase

Once a chromosomal population has been generated and a disease model is specified or generated, family-based or case/control datasets can be generated. Individuals are created by sampling chromosome pairs (one pair for each chromosomal group, i.e., autosome) from the population without replacement. When generating case/control datasets, individuals have a penetrance function applied to produce a case or control status. An individual's genotype is matched to a value from the penetrance function in the disease model, and a draw from a random number generator is used with this probability to assign affected or unaffected status. Individuals are drawn from the population and assigned status in this manner until the desired number of cases and controls is reached. When generating family-based datasets, two parents are generated by drawing chromosomes from the population (just as with individuals in case/control data), and within each parent, recombination occurs between the two chromosomes to form two gametes. Of these two gametes, one is selected from each parent, producing the two chromosomes in the offspring. This mating procedure is continued to produce the desired pedigree structure. Each individual in the pedigree is assigned affection status using the same procedure as in case/control data, and this process is repeated until the desired number of pedigrees with the correct structure and pattern of affection status is attained. GenomeSIMLA can generate multiple case/control and family-based datasets from the same chromosomal population, allowing simulation studies to account for sampling variation in method evaluation. All randomized processes in the simulation are governed by a single random seed specified by the user, so that simulations are reproducible.

D. Simulating additional types of noise in genomeSIMLA

Four additional types of error/noise can be introduced into simulated datasets: genotyping error, missing data, phenocopy, and genetic heterogeneity. Genotyping error, or the mislabeling of genotypes by calling algorithms, is simulated by randomly perturbing a user-specified percentage of genotype states. A SNP for an individual is randomly chosen, and the additive encoding for the genotype at that SNP (0, 1, 2) is either incremented or decremented according to a random draw (a fifty–fifty chance). Missing data is simulated by randomly deleting a user-specified percentage of the genotypes from the dataset. Phenocopy occurs when an individual acquires a phenotype from a nongenetic cause. A commonly used example is epilepsy, where strongly inherited forms are characteristically indistinguishable from epilepsy induced from head trauma—forms caused by trauma have no genetic basis and will reduce statistical power to detect the loci at play in the genetic form. Phenocopy is simulated by removing a user-specified percentage of cases from a drawn dataset and replacing those individuals with newly drawn individuals from the population who are assigned affected status regardless of their genotype. With family-based data, entire pedigrees are redrawn, keeping the same number of affected but assigning one or more as phenocopies. Genetic heterogeneity is the presence of multiple genetic models that lead to the same or similar disease state. In this case, genomeSIMLA has the ability to generate some predefined proportion of the cases under one genetic model and another proportion of the cases under a different genetic model. This type of error has been known to be extremely challenging for many data analysis methods (Edwards *et al.*, 2009; Ritchie *et al.*, 2003).

Case/control datasets are produced with additive encoded genotypes, in a space-delimited text file where the first column is affection status. Family-based datasets are produced in PED file format with a six-column header.

E. Simulation reporting

The forward-time population simulation phase produces several report files to characterize both the chromosomal pool representing the total population, and drawn datasets representing samples from that population. Report files are written in HTML format for multi-platform viewing.

1. Simulation overview

This report shows a chromosome-wide Haploview-style D′ plot for each chromosome, providing a general single glance sanity check that the simulation produced patterns of LD that are reasonable. This report also provides basic statistics, such as the number of alleles that fixed in the simulation, the final population size, and the number of called haplotype blocks.

2. Detailed block report

This report is produced for each chromosome in the simulation, and provides a chromosome-wide Haploview-style D′ and r^2 plot that is annotated with called haplotype blocks, the physical position of each marker, and the allele frequency of each marker. Haplotype block details are also shown in this report, with a higher resolution Haploview-style plot of LD in both D' and r^2 with the called block, bounding SNPs of the block, and physical position and allele frequency of markers in the block annotated.

3. Linkage disequilibrium report

This is a text file containing pairwise LD statistics, formatted like Haploview test output. This file includes D′, r^2, and LOD calculations.

4. Growth rate report

This report is produced once for the entire simulation, and shows a plot of the growth function, showing changes in population size over generations. This report also contains exact population sizes, given every 50 generations.

V. GENOMESIMLA: AN EXPERIMENTAL EXAMPLE

To develop an understanding of the consequences of different population growth curve parameter settings, a series of experiments were designed (Table 1.2). The hypothesis is that some combination of population growth parameters will emulate the average profile of correlation by distance observed in the HapMap data. We used a generalized logistic curve, or Richards curve, to model realistic population growth (Richards, 1959; Eq. (1.1)). The Richards growth curve consists of five parameters: A—the initial population size or lower asymptote,

Table 1.2. Parameter Sweep of Population Growth Parameters for the Logistic Function: Settings for Three Scans

Parameters	Scan 1	Scan 2	Scan 3
A—lower asymptote	500, 750, 1000	100, 150, 200, 250, 300	750, 1000, 1250, 1500
C—upper asymptote	120k, 500k, 900k	110k, 120k	120k
M—maximum growth time	305, 315, 325, 335, 345, 355	350, 400, 450	500, 1000, 1500, 2000, 2500, 3000
B—growth rate	0.005, 0.0075, 0.01	0.018, 0.02, 0.022, 0.025	0.02, 0.025, 0.03, 0.035, 0.04
T—maximum growth position	0.1. 0.2, 0.3	0.1	0.1
Total parameters	486	120	120

C—the carrying capacity of the population or the upper asymptote, M—the generation of maximal growth, *B*—the growth rate, and *T*—a parameter that determines if the point of maximal growth occurs near the lower or upper asymptote.

$$Y = A + \frac{C}{\left(1 + Te^{-B(x-M)}\right)^{1/T}} \tag{1.2}$$

This function provides a parameterized theoretical basis for population growth, though real population growth likely has more stochastic variability. To allow variability in population growth, we implemented a jitter parameter that draws a random number from a uniform distribution over a range specified by the user and adds or subtracts that percentage from the population size predicted by the growth curve. For the purposes of the parameter exploration in this study, however, the jitter parameter was set to zero.

We scanned through a wide range of parameters to find population growth profiles providing suitable correlation among genetic variables for data simulation. Since there were five parameters to vary and many possible values for each, we were still limited to a small subset of the possible sets of growth parameters available. Prior to this study, we performed a number of parameter sweeps to evaluate ranges that were likely to yield interesting and realistic LD patterns (results not shown) in a population of 100,000 chromosomes. For this study, we split the parameter sweep into three scans. In total, 726 combinations of parameter settings were examined for average LD over distance.

We predict that a common usage of genomeSIMLA software will be to simulate case–control and family-based whole-genome association datasets containing 500,000–1,000,000 biallelic markers across the genome. These data could be used to evaluate the sampling properties of new or established association methods or techniques to characterize the genetic structure of populations. While genomeSIMLA can simulate data of this magnitude, for this study, we wanted to focus on a single chromosome. Thus, we simulated the 6031 chromosome 22 markers used on the popular Affymetrix 500K SNP Chip.

To visualize the results of each parameter combination, average R^2 by distance in kilobases was graphed for the simulated data and for the CEPH (Caucasian), Yoruba (African), and Chinese/Japanese HapMap populations. This representation captures global estimates of correlation by distance across the entire chromosome.

Parameter settings in Scan 1 did not yield LD which was comparable to HapMap samples. A trend was observed among the better fitting models that the parameters C and *T* always functioned best when set to 120k and 0.1, respectively. Scan 2 examined very small initial populations and more rapid growth to strengthen LD profiles through rapid genetic drift. These unfortunately also resulted in fixing many alleles. Scan 3 focused on larger initial populations, late maximum growth, and rapid growth. These simulations were the most successful and resulted in several curves which approximated LD in HapMap samples well. One such example is presented in Fig. 1.4.

While not a perfect fit to any population, the curve represents a good approximation of the correlation observed in the data. Of note is the fit in the shorter ranges, since short range LD is more related to the power to detect associations with disease susceptibility loci (Durrant *et al.*, 2004). A sample of the actual LD observed among the markers is presented in Fig. 1.5. The goal of this study was to obtain data which on average is similar to HapMap data. Since we initialized the chromosomes with random minor allele frequency and the measure R^2 is sensitive to this parameter, it is not expected that each intermarker correlation will be identical to the value calculated from the HapMap data. However, it can be seen here that the major features and regions of high and low correlation are captured. The growth curve in Fig. 1.2 and the LD shown in Fig. 1.3 were generated with the following parameters: $A=1500$, $C=120000$, $M=500$, $B=0.02$, $T=0.1$. D', an alternate measure of LD, was more difficult to

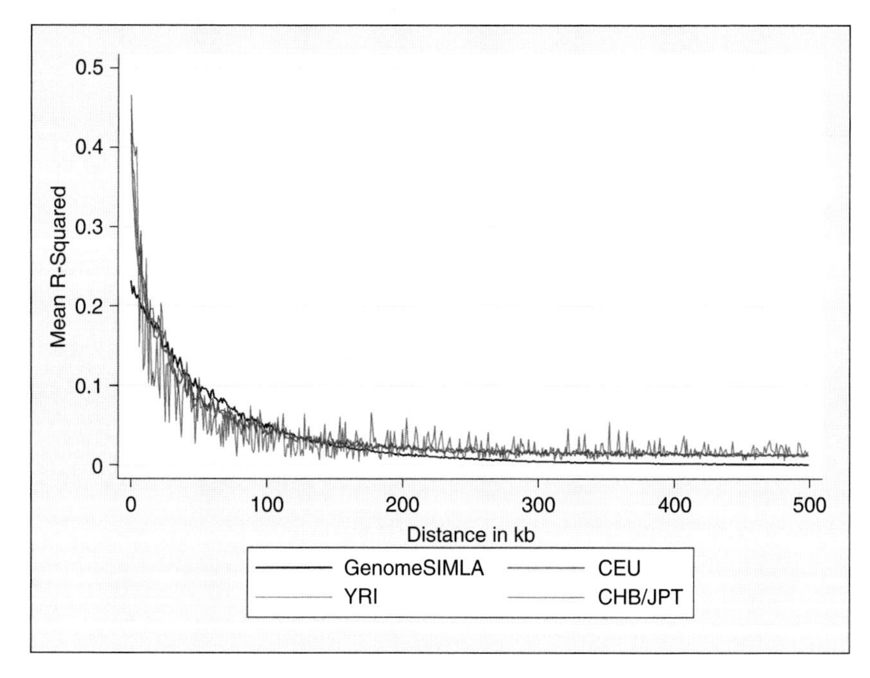

Figure 1.4. Average R^2 by distance (kb) for simulated, CEPH (Caucasian), YRI (Yoruba African), and CHB/JPT (Chinese/Japanese) samples.

fit than R^2. The curves for the simulated data generally were stronger than those observed for the real data in the short ranges but weaker at long ranges. The reasons for this are unknown but are a topic of further study for genomeSIMLA.

We also measured the time to completion for various size simulations. We examined the markers for the Affymetrix 500K in chromosomes 1 and 22 and the full chip (Table 1.3) for the growth parameters in Figs. 1.4 and 1.5.

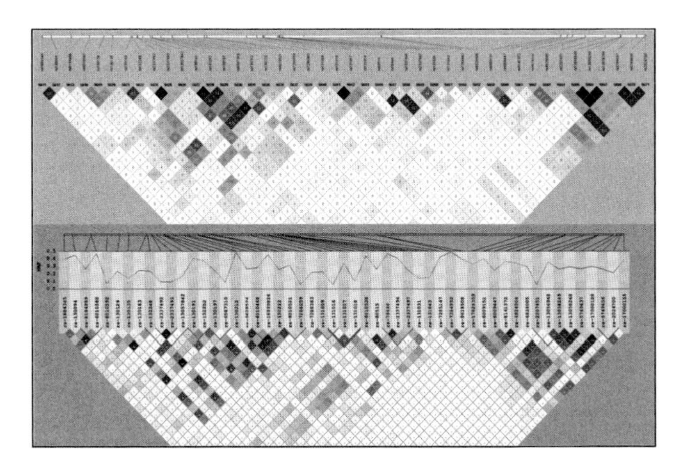

Figure 1.5. Sample of LD from the simulation detailed in Fig. 1.2 of R^2 plots from HapMap CEPH samples (above) and simulated data.

Table 1.3. Time to Completion for Pool Advancement to 100,000 Chromosomes and Graphical LD Calculation and Representation for up to 500,000 SNPs

Simulation	Processors	LD calculation	Time
Chr1	1	Sampled	13 min 41 s
Chr1	1	Complete	88 min 45 s
Chr1	4	Sampled	5 min 41 s
Chr1	4	Complete	33 min 4 s
Chr22	1	Sampled	2 min 15 s
Chr22	1	Complete	12 min 27 s
Chr22	4	Sampled	1 min 33 s
Chr22	4	Complete	4 min 30 s
500k	4	Sampled	74 min 52 s
500k	4	Complete	367 min 54 s
500k	8	Sampled	29 min 22 s
500k	8	Complete	123 min 21 s

In order to reduce the time required to scan a growth curve for ideal LD patterns, genomeSIMLA utilizes both sampled and complete LD. When generating sampled LD plots, genomeSIMLA draws LD plots for a small region (1000 SNPs) of each chromosome and limits the participants to a relatively small number (1500).

VI. CONCLUSIONS

In this chapter, we have reviewed a variety of algorithms and software tools available for the synthesis of genomic data for evaluation of novel computational and statistical methods. Four primary approaches have been developed for simulation genome data including family-based simulations, coalescent modeling, resampling methods, and forward population simulations. All of these approaches have strengths and weaknesses and the simulation tool of choice will depend heavily on the goals of the data simulation study.

A powerful simulation of genome-wide association study data can be conducted using genomeSIMLA (Edwards *et al.*, 2008). We found that tuning the parameters to emulate the average pattern of correlation in real human populations was difficult. However, some settings we used provided good qualitative fit to the observed real data. We initialized our chromosome pools with random allele frequency independent data, and only allowed the recombination probabilities to directly mimic those expected from the Kosambi function for the HapMap physical distances. This procedure was a proof of principle that it is neither necessary to directly resample HapMap chromosomes nor use computationally inefficient coalescent models to effectively simulate the properties of unobserved samples from real human populations, and that realistic data simulations can be achieved through forward population simulations. The speed and scale of the genomeSIMLA software is sufficient to provide timely results to investigators conducting power studies for various methods. The software architecture ensures that the user can access all available computational power to do very large whole-genome size studies. The time to completion for various size simulations for single and multiple processors are presented in Table 1.3. Those times include the time required to calculate and provide an interactive graphical interface of LD pictures for locus selection. These times are very fast given the computational task and represent the advanced implementation which is presented here. Demonstrations, manuals, and genomeSIMLA software for Mac, PC, and Linux are available for download at http://chgr.mc.vanderbilt.edu/genomeSIMLA. With this capability, researchers who develop novel methods to analyze genetic data can quickly and accurately estimate the performance and sampling properties of those methods.

While genomeSIMLA is a useful tool for the creation of *in silico* genome datasets, there is still further data simulation developments to be done. First, the existing data simulation software packages are limited in their ability to simulate truly admixed populations. Several existing tools can be modified to make this possible, but it is not a readily available option in the software. Second, while GWAS has been the workhorse of human genomics in recent years, next-generation sequencing is on the horizon. This means that more analytic methods will be developed for the analysis of rare variants as well as the combination of rare and common variants. Existing simulation tools can be extended for these types of data simulations, but again this is not an available option. Finally, many groups are interested in gene–environment interactions or systems biology approaches where data from SNPs or rare DNA variants will be combined with clinical/environmental data or data generated in transcriptome, proteome, or metabolome experiments. These types of quantitative independent variable data cannot currently be supported by most of the existing genome simulation software packages. Because this is a future area of research, many groups are considering extensions to existing genome simulation tools to accommodate these types of data.

Genome simulations are a critically important component to the evaluation of new statistical and computational methods for genetic analysis. These simulations are often criticized because they are much cleaner than natural, biological data. However, great strides have been made in the past several years to synthesize genome datasets *in silico* that better recapitulate natural, biological datasets. Additional advances are sure to follow and will certainly improve the complexity of the data simulations which will permit investigators to better assess new analytical methods.

References

Balloux, F. (2001). EASYPOP (version 1.7): A computer program for population genetics simulations. *J. Hered.* **92,** 301–302.

Bass, M. P., Martin, E. R., and Hauser, E. R. (1993). Pedigree generation for analysis of genetic linkage and association. *Pac. Symp. Biocomput.* **2004,** 93–103.

Boehnke, M. (1986). Estimating the power of a proposed linkage study: A practical computer simulation approach. *Am. J. Hum. Genet.* **39,** 513–527.

Boehnke, M., and Ploughman, L. M. (1997). SIMLINK: A Program for Estimating the Power of a Proposed Linkage Study by Computer Simulation. Version 4.12, April 2, 1997. http://csg.sph.umich.edu/boehnke/simlink.php

Carvajal-Rodriguez, A. (2008). GENOMEPOP: A program to simulate genomes in populations. *BMC Bioinform.* **9,** 223. http://www.biomedcentral.com/1471-2105/9/223.

Chen, G. K., Marjoram, P., and Wall, J. D. (2009). Fast and flexible simulation of DNA sequence data. *Genome Res.* **19,** 136–142.

Cook, N. R., Zee, R. Y., and Ridker, P. M. (2004). Tree and spline based association analysis of gene–gene interaction models for ischemic stroke. *Stat. Med.* **23,** 1439–1453.

Cristianini, N., and Shawe-Taylor, J. (2000). An Introduction to Support Vector Machines. Cambridge University Press, Cambridge.

Culverhouse, R., Suarez, B. K., Lin, J., and Reich, T. (2002). A perspective on epistasis: Limits of models displaying no main effect. *Am. J. Hum. Genet.* **70,** 461–471.

Culverhouse, R., Klein, T., and Shannon, W. (2004). Detecting epistatic interactions contributing to quantitative traits. *Genet. Epidemiol.* **27,** 141–152.

Dudek, S., Motsinger, A. A., Velez, D., Williams, S. M., and Ritchie, M. D. (2006). Data simulation software for whole-genome association and other studies in human genetics. *Pac. Symp. Biocomput.* 499–510.

Durrant, C., Zondervan, K. T., Cardon, L. R., Hunt, S., Deloukas, P., and Morris, A. P. (2004). Linkage disequilibrium mapping via cladistic analysis of single-nucleotide polymorphism haplotypes. *Am. J. Hum. Genet.* **75,** 35–43.

Edwards, T. L., Bush, W. S., Turner, S. D., Dudek, S. M., Torstenson, E. S., Schmidt, M., Martin, E., and Ritchie, M. D. (2008). Generating linkage disequilibrium patterns in data simulations using genomeSIMLA. *Lect. Notes Comput. Sci.* **4973,** 24–35.

Edwards, T. L., Lewis, K., Velez, D. R., Dudek, S., and Ritchie, M. D. (2009). Exploring the performance of multifactor dimensionality reduction in large scale SNP studies and in the presence of genetic heterogeneity among epistatic disease models. *Hum. Hered.* **67,** 183–192.

Ewing, G., and Hermisson, J. (2010). MSMS: A coalescent simulation program including recombination, demographic structure and selection at a single locus. *Bioinformatics* **26,** 2064–2065.

Frazer, K. A., Ballinger, D. G., Cox, D. R., Hinds, D. A., Stuve, L. L., Gibbs, R. A., Belmont, J. W., Boudreau, A., Hardenbol, P., Leal, S. M., Pasternak, S., Wheeler, D. A., *et al.* (2007). A second generation human haplotype map of over 3.1 million SNPs. *Nature* **449,** 851–861.

Hahn, L. W., Ritchie, M. D., and Moore, J. H. (2003). Multifactor dimensionality reduction software for detecting gene–gene and gene–environment interactions. *Bioinformatics* **19,** 376–382.

Hastie, T., Tibshirani, R., and Friedman, J. (2001). The Elements of Statistical Learning: Data Mining, Inference, and Prediction. Springer-Verlag, New York.

Hey (2005). A computer program for forward population genetic simulation.

Hoggart, C. J., Chadeau, M., Clark, T. G., Lampariello, R., De, I. M., Whittaker, J. C., and Balding, D. J. (2007). Sequence-level population simulations over large genomic regions. *Genetics* **177,** 1725–1731.

Hudson, R. R. (2002). Generating samples under a Wright–Fisher neutral model of genetic variation. *Bioinformatics* **18,** 337–338.

Kaufman, L., and Rousseeuw, P. J. (1990). Finding Groups In Data: An Introduction to Cluster Analysis Wiley-Interscience Publication, New York.

Kingman, J. (1982). The coalescent. *Stochastic Processes Appl.* **13,** 235–248.

Kooperberg, C., Ruczinski, I., LeBlanc, M. L., and Hsu, L. (2001). Sequence analysis using logic regression. *Genet. Epidemiol.* **21**(Suppl. 1), S626–S631.

Li, J., and Chen, Y. (2008). Generating samples for association studies based on HapMap data. *BMC Bioinform.* **9,** 44.

Li, C., and Li, M. (2008). GWAsimulator: A rapid whole-genome simulation program. *Bioinformatics* **24,** 140–142.

Liang, L., Zollner, S., and Abecasis, G. R. (2007). GENOME: A rapid coalescent-based whole genome simulator. *Bioinformatics* **23,** 1565–1567.

Mailund, T., Schierup, M. H., Pedersen, C. N., Mechlenborg, P. J., Madsen, J. N., and Schauser, L. (2005). CoaSim: A flexible environment for simulating genetic data under coalescent models. *BMC Bioinform.* **6,** 252.

Marchini, J., Donnelly, P., and Cardon, L. R. (2005). Genome-wide strategies for detecting multiple loci that influence complex diseases. *Nat. Genet.* **37,** 413–417.

Marjoram, P., and Wall, J. D. (2006). Fast "coalescent" simulation. *BMC Genet.* **7,** 16.

Moore, J. H., Hahn, L. W., Ritchie, M. D., Thornton, T. A., and White, B. (2004). Routine discovery of high-order epistasis models for computational studies in human genetics. *Appl. Soft Comput.* **4,** 79–86.

Motsinger, A. A., Ritchie, M. D., and Reif, D. M. (2007). Novel methods for detecting epistasis in pharmacogenomics studies. *Pharmacogenomics* 1229–1241.

Nelson, M., Kardia, S., Ferrell, R., and Sing, C. (2001). A combinatorial partitioning approach to identify multilocus genotypic partitions that predict quantitative trait variation. *Genome Res.* (proof1–13).

Ott, J. (1989). Computer-simulation methods in human linkage analysis. *Proc. Natl Acad. Sci. USA* **86,** 4175–4178.

Peng, B., and Kimmel, M. (2005). simuPOP: A forward-time population genetics simulation environment. *Bioinformatics* .

Peng, B., Amos, C. I., and Kimmel, M. (2007). Forward-time simulations of human populations with complex diseases. *PLoS Genet.* **3,** e47.

Ploughman, L. M., and Boehnke, M. (1989). Estimating the power of a proposed linkage study for a complex genetic trait. *Am. J. Hum. Genet.* **44,** 543–551.

Richards, F. (1959). A flexible growth function for empirical use. *J. Exp. Bot.* **10,** 290–300.

Ripley, B. D. (1996). Pattern Recognition and Neural Networks Cambridge University Press, .

Ritchie, M. D., Hahn, L. W., Roodi, N., Bailey, L. R., Dupont, W. D., Parl, F. F., and Moore, J. H. (2001). Multifactor-dimensionality reduction reveals high-order interactions among estrogen-metabolism genes in sporadic breast cancer. *Am. J. Hum. Genet.* **69,** 138–147.

Ritchie, M. D., Hahn, L. W., and Moore, J. H. (2003). Power of multifactor dimensionality reduction for detecting gene–gene interactions in the presence of genotyping error, missing data, phenocopy, and genetic heterogeneity. *Genet. Epidemiol.* **24,** 150–157.

Schmidt, M., Hauser, E. R., Martin, E. R., and Schmidt, S. (2004a). Extension of the SIMLA package for generating pedigrees with complex inheritance patterns: Environmental covariates, gene–gene and gene–environment interaction. *Stat. Appl. Genet. Mol. Biol.* **2005** (Article 15).

Schmidt, M., Hauser, E. R., Martin, E. R., and Schmidt, S. (2004b). Extension of the SIMLA package for generating pedigrees with complex inheritance patterns: Environmental covariates, gene–gene and gene–environment interaction. *Stat. Appl. Genet. Mol. Biol.* **2005** (Article 15).

Spencer, C. C., and Coop, G. (2004). SelSim: A program to simulate population genetic data with natural selection and recombination. *Bioinformatics* **20,** 3673–3675.

Terwilliger, J. D., and Ott, J. (1994). Handbook of Human Genetic Linkage. Johns Hopkins University Press, Baltimore.

Terwilliger, J. D., Speer, M., and Ott, J. (1993). Chromosome-based method for rapid computer simulation in human genetic linkage analysis. *Genet. Epidemiol.* **10,** 217–224.

Weeks, D. E., Ott, J., and Lathrop, G. M. (1990). SLINK: A general simulation paorgram for linkage analysis. *Am. J. Hum. Genet.* **47,** A204.

Wille, A., Hoh, J., and Ott, J. (2003). Sum statistics for the joint detection of multiple disease loci in case–control association studies with SNP markers. *Genet. Epidemiol.* **25,** 350–359.

Wolfram, S. (1994). Cellular Automata and Complexity. Addison-Wesley Publishing Company, Reading.

Wright, F. A., Huang, H., Guan, X., Gamiel, K., Jeffries, C., Barry, W. T., de Villena, F. P., Sullivan, P. F., Wilhelmsen, K. C., and Zou, F. (2007a). Simulating association studies: A data-based resampling method for candidate regions or whole genome scans. *Bioinformatics* **23,** 2581–2588.

Wright, F. A., Huang, H., Guan, X., Gamiel, K., Jeffries, C., Barry, W. T., de Villena, F. P., Sullivan, P. F., Wilhelmsen, K. C., and Zou, F. (2007b). Simulating association studies: A data-based resampling method for candidate regions or whole genome scans. *Bioinformatics* **23,** 2581–2588.

Zhang, F., Liu, J., Chen, J., and Deng, H. W. (2008). HAPSIMU: A genetic simulation platform for population-based association studies. *BMC Bioinform.* **9,** 331.

Zhu, J., and Hastie, T. (2004). Classification of gene microarrays by penalized logistic regression. *Biostatistics* **5,** 427–443.

2

Logic Regression and Its Extensions

Holger Schwender and Ingo Ruczinski
Department of Biostatistics, Bloomberg School of Public Health, Johns Hopkins University, Baltimore, Maryland, USA

ABSTRACT

Logic regression is an adaptive classification and regression procedure, initially developed to reveal interacting single nucleotide polymorphisms (SNPs) in genetic association studies. In general, this approach can be used in any setting with binary predictors, when the interaction of these covariates is of primary interest. Logic regression searches for Boolean (logic) combinations of binary

Advances in Genetics, Vol. 72

0065-2660/10 $35.00
DOI: 10.1016/S0065-2660(10)72002-2

variables that best explain the variability in the outcome variable, and thus, reveals variables and interactions that are associated with the response and/or have predictive capabilities. The logic expressions are embedded in a generalized linear regression framework, and thus, logic regression can handle a variety of outcome types, such as binary responses in case-control studies, numeric responses, and time-to-event data. In this chapter, we provide an introduction to the logic regression methodology, list some applications in public health and medicine, and summarize some of the direct extensions and modifications of logic regression that have been proposed in the literature.

I. INTRODUCTION

In the analysis of genotype data, the effect sizes of associations between individual SNPs (single-nucleotide polymorphisms) and a response of interest (e.g., the disease status in a case-control studies, or a quantitative trait such as height in a population-based study) are typically small. Therefore, it is assumed that not only single SNPs, but also interactions of several SNPs contribute to the variation in the phenotype, for example, the disease risk (Culverhouse *et al.*, 2002; Garte, 2001). Two-way interactions can be exhaustively investigated even in genome-wide association studies (Marchini *et al.*, 2005); the search for higher order interactions, however, is currently limited to smaller sets of markers, such as sets of tagging SNPs or the markers from a candidate gene study.

A higher order interaction could, for example, be defined by the following rule:

> ***IF*** SNP S_1 is of the homozygous reference genotype **AND** SNP S_2 is of the homozygous variant genotype **OR** both SNP S_3 **AND** S_4 are **NOT** of the homozygous reference genotype,
> ***THEN*** a person has a higher risk to develop a particular disease.

In particular, these Boolean rules encode many common genetic models that involve epistatic interactions, for example, the following double-penetrance model:

> ***IF*** simultaneously SNP S_1 has two variant alleles **AND** SNP S_2 has at least one variant allele **OR** SNP S_2 has two variant alleles **AND** SNP S_1 has at least one variant allele,
> ***THEN*** a person has a higher disease risk.

In other words, a subject with at least three variant alleles across the two SNPs is at higher risk for disease (Fig. 2.1).

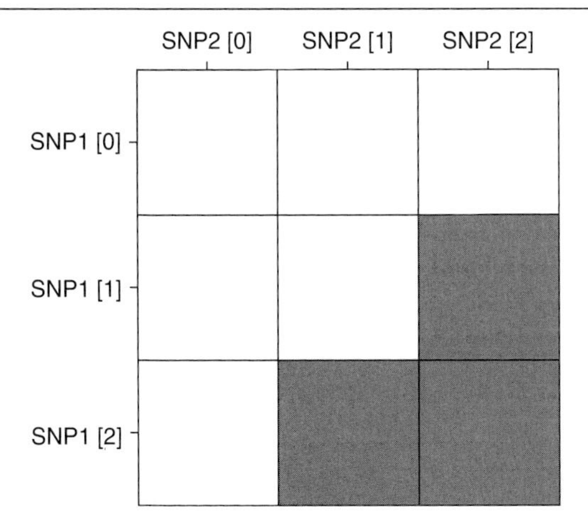

Figure 2.1. An example of a genetic "double penetrance" model. Subjects with two variant alleles at one marker and at least one variant allele at another marker (indicated by the grey region) have a higher disease risk.

To describe these genetic models, it is convenient to encode the information for SNPs (factors with levels 0, 1, 2, specifying the number of variant alleles) as binary predictors using dominant and recessive coding. That is, we are searching for AND- and OR-combinations of statements such as

S_{iD}: "At least one of the chromosomes contains the minor allele at locus S_i,"
S_{iR}: "Both chromosomes contain the minor allele at locus S_i."

These statements are either TRUE or FALSE. Mathematically speaking, we are interested in detecting Boolean combinations of logic variables associated with the response. Using the above-mentioned variables, which code for dominant and recessive genetic effects, the two Boolean (or logic) expressions $(S_{1D}{}^C \wedge S_{2R}) \vee (S_{3D} \wedge S_{4D})$ and $(S_{1R} \wedge S_{2D}) \vee (S_{2R} \wedge S_{1D})$ underlie the two ***IF-THEN*** statements given above. Here, $\wedge$ denotes the Boolean AND-combination, $\vee$ denotes the Boolean OR-combination, and the complement of any variable is indicated by the Boolean complement ^{C}. For example, $S_{1D}{}^C$ is equal to one (TRUE) if SNP S_1 does *not* have any variant alleles.

A method specifically developed to identify Boolean combinations of binary covariates as predictors for an outcome is logic regression (Ruczinski *et al.*, 2003). This adaptive classification and regression procedure is not restricted to SNP data, but can be applied to any type of data based on binary variables. It can be used to classify records with binary outcome data, but also in a regression framework. By embedding Boolean expressions into a generalized linear model framework, logic regression allows for the combination of several logic expressions into a single model, and the inclusion of additional covariates as main effects. Any type of response (binary, continuous, time-to-event) can be investigated, as long as an objective function can be defined. This approach also allows for the assessment of the magnitude of the disease risk, for example:

> ***IF*** a subject exhibits at least one minor allele at both S_1 **AND** S_2,
> ***THEN*** the odds of having a certain disease are three times larger for this person than for a subject of the same age showing the homozygous reference genotype at one or both loci.

Logic regression should not be confused with logistic regression, a member of the generalized linear model family suitable for data with a binary response (i.e. a binary outcome). While SNP–SNP interactions can be assessed using logistic regression (e.g., in case-control studies), these interactions need to be known in advance, and used as input variables in the model. By contrast, logic regression is applicable for any type of response, as long as the predictors (i.e., the independent variables) are binary. Interactions of interest need not be known in advance, quite the contrary, the detection of important variable interactions is the main aim of logic regression. The actual binary variables are merely the input for this procedure.

It should be noted that similar to logistic regression, logic regression can also be employed to predict the outcomes for new observations. However, it is usually the case that markers in SNP association studies only exhibit small effect sizes, even when considered simultaneously, and thus have very little predictive power, leading to poor sensitivity and specificity, for example, for subject-specific disease prognosis. Thus, we only focus on association and feature selection, that is, the detection of disease-associated SNPs and SNP–SNP interactions, the primary objective of genetic association studies. We also note that logic regression is only applicable to candidate gene studies that include at most a few thousand SNPs, or to a subset of markers selected from genome-wide association studies. As described in more detail below, the search space of the Boolean models investigated is immense, with many local optima, and thus, logic regression is primarily based on a probabilistic search algorithm (although a greedy option has also been implemented).

In the following, we give a detailed description of the logic regression algorithm (Sections II.A–II.C). While originally devised and successfully used in the context of genetic association studies (see, e.g., Kooperberg *et al.*, 2001; Ruczinski *et al.*, 2004; Witte and Fijal, 2001), logic regression has also been employed in a variety of other Biomedical research settings. A brief overview of some of these applications is given (Section II.D). Further, several important methodological modifications and extensions of logic regression have been proposed and implemented. These are described in an overview at the end of this chapter (Section III).

II. LOGIC REGRESSION

Logic regression can be employed as a classification tool, but can also be used to derive a regression model with Boolean terms as covariates and the respective parameter estimates. In the classification approach, we aim to find a single Boolean expression that predicts a binary outcome. In the regression framework, several Boolean expressions can be investigated and simultaneously embedded in a (generalized) linear regression model. In other words, the Boolean combinations of the binary predictors themselves form the covariates used in the model, related to the outcome by a link function.

Logic Regression is available as Free Software under the terms of the Free Software Foundation's GNU General Public License in source code form. The software has been implemented in the statistical software environment `R` (R Development Core Team, 2009), and can be downloaded as package `LogicReg` from the The Comprehensive R Archive Network (http://cran.r-project.org/).

A. Logic trees

Logic regression is an adaptive discrimination methodology based on Boolean combinations of logic variables, that is, binary variables that are either TRUE or FALSE. These variables can be negated by the operator ^{C} (e.g., the term $S_{iR}{}^C$ represents the statement "SNP S_i is NOT of the homozygous variant genotype"), and combined by the operators $\wedge$ (AND) and $\vee$ (OR) to form a logic expression L, that in turn is also either TRUE or FALSE. In the logic regression algorithm, Boolean expressions such as $(S_{1D} \wedge S_{2D}{}^C) \vee S_{3R}$ are represented by so-called "logic trees" (Fig. 2.2). The nodes of a logic tree consist of the AND- and OR-operators, and the leaves (i.e., the terminal nodes) show the variables, where the complement (negation) of a binary variable is shown by white letters on a black background. If in a case-control study the above logic expression is used as classification rule, then a subject will be classified as case if L is TRUE, which is the case when either both S_{1D} and $S_{2D}{}^C$ are TRUE (i.e., S_{1D} is TRUE, and S_{2D} is FALSE), or S_{3R} is TRUE (or both $S_{1D} \wedge S_{2D}{}^C$ and S_{3R} are TRUE, since $\vee$ is *not* an exclusive OR). Otherwise, the subject is classified as control (Fig. 2.2).

It is noteworthy that while logic trees and classification and regression trees (CART; Breiman *et al.*, 1984) share some similarities, they are very different constructs. Each Boolean term as represented by a logic tree can be transformed into a classification and regression tree, and vice versa. In CART trees, however, the nodes comprise the variables (and the split based on them), and the leaves contain the predictions for new observations based on these decision trees. Thus, the tree is evaluated in a top-down fashion. Logic trees, on the other hand, are evaluated in a bottom-up fashion, where the variables are encoded in the leaves, and the prediction for a subject is the outcome of the evaluation of the Boolean statement that appears in the root (top) node of the tree (Fig. 2.2).

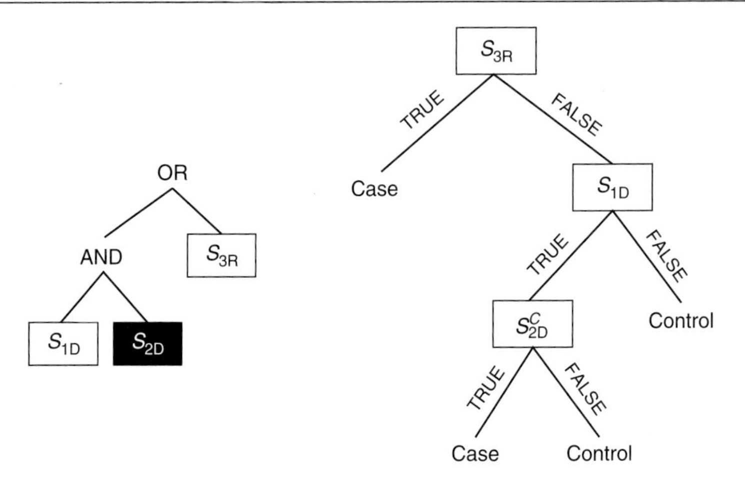

Figure 2.2. Logic tree versus CART tree: the same classification rule based on the binary variables S_{1D}, S_{2D}, and S_{3R} is shown as logic tree (left), and as classification and regression tree (right).

B. Model search

Even with a modest number of genetic markers, the search space defined by the set of all Boolean combinations is vast, and thus, an efficient search is imperative. The search strategy is defined by the move set that determines the "neighborhood" of similar logic models, the actual search algorithm, and the objective function that quantifies and compares logic models. These parts are discussed in the following subsections.

1. The move set

Logic trees are graphical representations of Boolean expressions, but also build the basis for the set of moves employed in the greedy or probabilistic search algorithm underlying logic regression. These moves generate new logic trees in the search for the best logic regression model as assessed by an objective function (such as the misclassification rate in a classification problem). There are six permissible moves in this tree-growing process (Fig. 2.3). In each step of the search algorithm, we can:

- *Alternate a leaf.* A literal, that is, a logic variable or its complement, is replaced by another literal, where replacements that create tautologies in the tree are

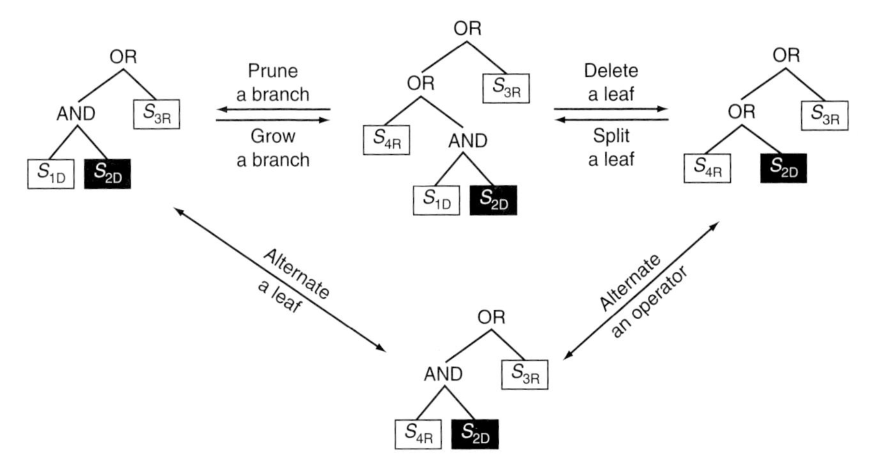

Figure 2.3. The move set of the logic regression algorithm, as implemented in the publicly available software.

not permitted. For example, $S_{2D}{}^{C}$ can be replaced by any literal except for S_{1D} and $S_{1D}{}^{C}$ in the left-most logic tree displayed in Fig. 2.3.

- *Alternate an operator*. An OR is replaced by an AND, or vice versa.
- *Grow a branch*. At each node (except for the terminal nodes, i.e., the leaves), a new branch can be grown by cutting the subtree starting at this node, adding a new operator to this node, and adding the subtree to the right branch and a leaf to the left branch of this new node. In Fig. 2.3, for example, the subtree $S_{1D} \wedge S_{2D}{}^{C}$ is cut from the left-most tree, and an OR is added to the corresponding node. Then, S_{4R} and $S_{1D} \wedge S_{2D}{}^{C}$ are added as the children to this node to generate the tree in the upper middle of Fig. 2.3.
- *Prune a branch*. This is the countermove to growing a branch, in which a branch is removed from the tree by selecting a node that is a parent node to a leaf and an operator, and removing this node and the corresponding leaf. Thus, only the two OR-nodes in the logic tree in the upper middle of Fig. 2.3 can be pruned, as the children of the AND-node are both leaves (namely S_{1D} and $S_{2D}{}^{C}$).
- *Split a leaf*. Each leaf can be split by replacing it with a subtree, composed of an operator, the split leaf itself, and another leaf. In the right-most tree of Fig. 2.3, for example, $S_{2D}{}^{C}$ is split by replacing it with $S_{1D} \wedge S_{2D}{}^{C}$, to generate the tree in the upper middle of this figure.
- *Delete a leaf*. This is the countermove to splitting a leaf. A leaf, whose sibling is also a leaf, is deleted by removing it from the tree and replacing the operator in its parent node by its sibling. In the logic tree in the upper middle of Fig. 2.3,

this move can thus only be applied to S_{1D} and S_{2D}^{C} (as the siblings of both S_{3R} and S_{4R} are operators).

A move proposed to investigate a new logic model is selected from the list of all permissible moves, which might not include the entire list of moves above. For example, if a logic model of a certain size (defined by the total number of binary variables in the logic trees) is desired, and the model under current evaluation is already of that desired size, any move that increases the model size (splitting a leaf, growing a branch) is not permitted.

2. The search strategy

Logic regression can be carried out using a greedy search algorithm. More commonly however, simulated annealing is employed to find good scoring models. In a greedy search (as used, e.g., in CART), all possible moves are considered, and the one that optimizes the objective function (misclassification rate, deviance, residual sum of squares, etc.) among those moves is chosen, provided that the model fitness improves. Otherwise, the search is terminated. The main advantage of this approach is the low computational burden. The main drawback, on the other hand, is that such a greedy search often only finds a model that is locally optimal (i.e., in the vicinity of the currently considered models), but does not provide the best possible solution among all possible models in the search space. As the logic regression models define a very complex search space with a vast number of local optima, the search for the best scoring model is usually probabilistic, that is, based on simulated annealing (Kirkpatrick *et al.*, 1983). In each iteration of this simulated annealing algorithm, a new logic tree is proposed by randomly executing a move from the collection of permissible moves listed above. The acceptance probability for the new tree (i.e., the probability for replacing the current model) is based on the objective function comparing the performances of the two models, and a simulated annealing-specific parameter called temperature. The acceptance probability is always equal to one if the new tree exhibits a better score than the current model. The acceptance probability for moves that worsen the score is positive but converges to zero as the annealing progresses. Thus, getting stuck in local optima can be avoided in the search for the global optimum.

We note that in the logic regression move set each move has a "countermove," that is, the reversal of a move is also permissible. Further, any logic expression (tree) can be reached from any other logic expression (tree) in a finite number of moves without periodicities. This defines an irreducible and aperiodic (and thus, ergodic) Markov chain, a very desirable property to assure convergence in simulated annealing-based search algorithms (see Kirkpatrick *et al.*, 1983; Ruczinski, 2000).

3. Objective functions

Logic regression can be employed as a classification tool, but also used in a regression framework. We first describe the classification approach, which aims to find a single Boolean expression that predicts a binary outcome. In the regression framework, several Boolean expressions can be investigated and simultaneously embedded in a generalized linear regression model. In other words, the Boolean combinations of the binary predictors themselves form the covariates used in the model, related to the outcome by a link function.

The classification approach of logic regression is summarized in Algorithm 1. It is an adaptive discrimination methodology for predicting the binary response of interest based on Boolean combinations of logic variables, that is, of binary variables that are either TRUE or FALSE.

Algorithm 1: Classification Approach of Logic Regression

Assume that the following is given: a data set, an annealing cooling scheme, the maximum number n_{leaf} of leaves permitted in the Boolean term, and the number n_{iter} of iterations to be carried out.

1. Initially, specify L_0 by randomly drawing a logic tree consisting of one leaf.
2. For $k = 1, \ldots, n_{\text{iter}}$,
   (a) propose a new tree L_{new} by randomly selecting a permissible move, and the variable and/or operator involved in this move (if applicable),
   (b) accept the new tree with probability

$$a(M_{k-1}, M_{\text{new}}, T) = \min\{1, \exp((M_{k-1} - M_{\text{new}})/T)\}$$

as logic regression model L_k, where M_{k-1} and M_{new} are the misclassification error rates of L_{k-1} and L_{new}, respectively, and the temperature T is specified by the simulated annealing cooling scheme.

A typical cooling scheme starts at $T = 10^{z_{\text{start}}}$, and then the temperature is lowered to $T = 10^{z_{\text{end}}}$, $z_{\text{end}} < z_{\text{start}}$, in equal decrements on the $\log_{10}$-scale. Thus, the acceptance probability at the very beginning of the annealing scheme is in essence equal to one for all models proposed, and thus, many Boolean terms are investigated. As the temperature decreases, so do the acceptance probabilities for worse scoring models, and at the end of the annealing scheme almost no new tree is accepted if its misclassification rate is larger than the misclassification rate of the current tree.

Table 2.1 shows an example of such a cooling scheme, in which the misclassification rate of the proposed new logic tree is 0.05 units larger than the misclassification rate of the current logic regression tree, which determines the acceptance probability $a(M_{k-1}, M_{\text{new}}, T) = \exp(-0.05/T)$. The table shows that in the first 100,000 iterations the new tree is accepted with a probability of 99.5%, although its misclassification rate is worse than the misclassification rate

Table 2.1. Example for a Cooling Scheme with $z_{start} = 1$, $z_{end} = -3$, $n_{iter} = 500{,}000$, $M_{k-1} = 0.20$, and $M_{new} = 0.25$

Iterations	T	$a(M_{k-1}, M_{new}, T)$
1–100,000	10^1	0.9950
100,001–200,000	10^0	0.9512
200,001–300,000	10^{-1}	0.6065
300,001–400,000	10^{-2}	0.0067
400,001–500,000	10^{-3}	1.93×10^{-22}

The best cooling scheme highly depends on the actual data sets.

of the current model. This acceptance probability becomes rapidly smaller as the temperature is decreased, and the newly proposed model is virtually never accepted in the last 100,000 iterations of the annealing scheme.

The above is just one simple example for a possible cooling scheme. When an annealing scheme is devised, several criteria (such as the total number of predictors available) need to be considered. Unfortunately, this is more of an art than a science, and there are no hard and fast rules how to find the best possible annealing algorithm. The logic regression software provides a feature for monitoring the annealing process. Upon an initial scan through the data, it also suggests values for z_{start}, z_{end}, and n_{iter} if requested by the user. However, a carefully (and manually) devised annealing scheme typically yields better results.

In addition to the above-described classification framework, logic regression can also be employed in regression settings (as the name implies). Several types of link functions have been implemented in the software, and thus, linear regression, logistic regression, Cox proportional hazard models, and exponential survival models are readily available. In fact, any type of model can be considered as long as an objective or scoring function can be defined (such as the residual sum of squares, the deviance, the negative partial likelihood, and the negative log-likelihood for the above-mentioned model types). This regression approach enables the inclusion of multiple Boolean expressions L_j, $j = 1, \ldots, p$, into a generalized linear model framework

$$g(E(Y)) = \beta_0 + \sum_{j=1}^{p} \beta_j L_j.$$

Here, Y is the response, β_j, $j = 0, \ldots, p$, are the parameters, and g is a link function. If Y is a binary variable, for example, encoding case-control status, then the logit function is used in logic regression. If Y is quantitative, $g()$ is the linear link. Additional covariates X_a, $a = 1, \ldots, A$, such as age or sex, which need not be binary, can also be included in the model. In this case, the above becomes

$$g(E(Y)) = \beta_0 + \sum_{a=1}^{A} \gamma_a X_a \sum_{j=1}^{p} \beta_j L_j.$$

In this instance, the variables X_a are always included in the logic regression model. Since in the regression setting the search also starts with a single tree consisting of a single, randomly drawn literal, a seventh move needs to be added to the move set displayed in Fig. 2.3, which adds a new tree consisting of a randomly selected literal to the model. This move is permissible as long as the number p of logic trees is smaller than the user-specified maximum number n_{tree} of trees. The regression approach of logic regression is summarized in Algorithm 2.

Algorithm 2: Regression Approach of Logic Regression

Assume that the following is given: a data set, an annealing cooling scheme, a link function g, an objective function f, the maximum number n_{leaf} of leaves, the maximum number n_{tree} of trees, and the number n_{iter} of iterations used in the annealing.

1. Initially, specify L_0 by randomly drawing a logic tree consisting of one leaf.
2. For $k = 1, \ldots, n_{\text{iter}}$,
   (a) either randomly choose one of the logic trees and apply a randomly selected move to this tree, or—if $p < n_{\text{tree}}$—add a new tree consisting of a random literal to the model,
   (b) fit the logic regression model (including additional covariates, if applicable),
   (c) accept the new logic regression model with probability

$$a(\hat{y}_{k-1}, \hat{y}_{\text{new}}, y, T) = \min\{1, \exp((f(\hat{y}_{k-1}, y) - f(\hat{y}_{\text{new}}, y))/T)\},$$

where $f(\hat{y}_{k-1}, y)$ and $f(\hat{y}_{\text{new}}, y)$ are the values of the objective function for the current and the new logic regression model, respectively, and the temperature T is specified by the cooling scheme.

C. Model selection

To avoid overfitting in logic regression models, a selection procedure has to be carried out that assesses the "signal" in the data, and rejects models that include potentially spurious covariates. In this model selection step, a definition of model size is required. There are many reasonable definitions for model size. In logic regression, model size by default is defined as the combined total number of variables in all Boolean terms of the logic regression model, or equivalently, the total number of leaves in the trees representing the Boolean term(s). In the model search, candidate models of various sizes and various numbers of Boolean terms are generated in separate annealing chains. This is achieved by prohibiting moves that result in model sizes beyond the allowed complexity. If prediction is the main objective, cross-validation is the chosen method for model selection. Alternatively, when sufficient data are available, a training and test set approach can also be employed. In many studies, for example, SNP association studies, prediction is not the objective, but the detection of covariates associated with the response. In these settings, sequential permutation tests are employed (see Ruczinski *et al.*, 2003 for details) to determine the optimal model size. In practice, we find that the logic models chosen by cross-validation or permutation tests rarely exceed sizes of four or five. We note that the algorithms to perform cross-validation and permutation tests are available as options in the logic regression software package.

D. Applications in public health and medicine

Logic regression has been applied in numerous SNP association studies with a wide variety of phenotypes. In the context of oncologic studies, findings of higher order SNP interactions have been reported, for example, for prostate cancer (Etzioni *et al.*, 2004), breast cancer (Justenhoven *et al.*, 2008), colorectal cancer (Janes *et al.*, 2005; Poole *et al.*, 2010; Suchiro *et al.*, 2008), bladder cancer (Andrew *et al.*, 2008), and head and neck squamous-cell cancer (Harth *et al.*, 2008; Ickstadt *et al.*, 2008). Other noncancer phenotypes interrogated include hypertension (Huang *et al.*, 2004), myocardial infarction and ischemic stroke (Enquobahrie *et al.*, 2008; Kooperberg *et al.*, 2007), type 2 diabetes (An *et al.*, 2009), and cardiovascular disease (Carty *et al.*, 2010). An extension of logic regression called Monte Carlo logic regression (see Section III.C) has, for example, been applied to a schizophrenia study by Nicodemus *et al.* (2010).

Logic regression has also been applied in the context of other biomedical research. For example, Keles *et al.* (2004) combine a method for identifying transcription factor binding sites with logic regression to detect regulatory motifs. Segal *et al.* (2004) use logic regression and other procedures to predict HIV-1 replication capacity based on amino acid sequences. Feng *et al.* (2005) apply logic regression to DNA methylation data to identify combinations of hypermethylated genes. Yaziji *et al.* (2006) analyze immunohistochemical data to find logic expressions based on antibodies best suited for distinguishing between malignant epithelioid mesothelioma and adenocarcinoma. Vaidya *et al.* (2008) employ logic regression to detect combinations of urinary biomarkers associated with acute kidney injury, and Eliot *et al.* (2009) analyze flow cytometry data to identify immulogical factor combinations as predictors for the CD4 T cell recovery in HIV-1 infected subjects.

III. MODIFICATIONS AND EXTENSIONS OF LOGIC REGRESSION

Recently, several important extensions and modifications of logic expression have been proposed. In this section, these are described and discussed in some detail. The original logic regression approach, applicable to settings with binary or numeric outcomes in population-based studies (e.g., case-control or cohort studies), has been extended to allow for the analysis of multicategorical responses (described in Section III.A), and case-parent trio data from family-based genetic studies (Section III.B). Further, alternatives for quantifying "information" derived by logic regression have been developed, for example, Monte Carlo logic regression (Section III.C) and logicFS (Section III.D), which consider the importance of covariates and their interactions, instead of deriving a single model. Alternatives to the simulated annealing-based search have been proposed, for example, genetic programming and evolutionary algorithms to search for disease-associated logic expressions (Section III.E), or to detect Boolean combinations of haplotypes (Section III.F). Further modifications and extensions exist such as a fully Bayesian logic regression (Fritsch and Ickstadt, 2007), a Bayesian logic regression assuming perfect phylogeny (Clark *et al.*, 2007), and ROC curves for logic rules generated by logic regression (Etzioni *et al.*, 2003). These approaches are important and should be mentioned, but are not discussed in more detail here.

A. Multinomial logic regression

Logic regression can be applied to data when the predictors are binary, and a variety of responses can be handled using the standard `R` function `logreg` contained in the `R` package `LogicReg`. This includes binary and quantitative responses,as well as time-to-event data that can be analyzed via proportional hazard and exponential survival models. The `R` function `mlogreg` in the `R` package `logicFS` available at the BioConductor website http://www.bioconductor.org enables the analysis of multicategorical responses. This extended functionality allows for the fit of a multinomial logic regression model, analogous to multinomial logistic regression (Hosmer and Lemeshow, 1989). In this multinomial logic regression, one of the C groups (typically, a control group) is specified as the reference class, and $C-1$ logic regression models with the logit link function are fitted, each comparing the reference group with one of the $C-1$ nonreference groups. The probability p_c that a subject belongs to the nonreference class $c = 1, \ldots, C-1$ is estimated by

$$\hat{p}_c = \frac{\exp(\hat{\beta}_{0c} + \sum_{j=1}^{p} \hat{\beta}_{jc}\ell_{jc})}{1 + \sum_{h=1}^{C-1} \exp(\hat{\beta}_{0h} + \sum_{j=1}^{p} \hat{\beta}_{jh}\ell_{jh})}$$

where $\beta_{jc}, j = 0, \ldots, p$, are the parameters in the logic regression model comparing class c with the reference group, and ℓ_{jc} are the observed values of L_{jc} for a subject. The estimated probability $\hat{p}_0$ for a subject to belong to the reference group is given by

$$\hat{p}_0 = \left(1 + \sum_{h=1}^{C-1} \exp(\hat{\beta}_{0h} + \sum_{j=1}^{p} \hat{\beta}_{jh}\ell_{jh})\right)^{-1}.$$

B. Trio logic regression

Initially, logic regression has been developed for population-based studies, that is, for samples of unrelated subjects. A first extension of logic regression for family-based data was proposed by Li *et al.* (2010), applicable to case-parent trio data. This extension is based on a case-pseudo-control approach similar to the genotypic transmission disequilibrium test (gTDT; Schaid, 1996, 1999; Cordell *et al.*, 2004), which uses the conditional logistic likelihood. For a single marker, trio data can be set up for conditional logistic regression with a 1:3 matching ratio for the case genotypes versus three possible Mendelian realizations given the parents (the pseudo-controls). In logic regression however, all markers need to be considered simultaneously to identify the higher order interactions. This, for example, means that a total of $4^m - 1$ pseudo-controls exist per case when m unlinked SNPs are typed for the affected children and their parents. To circumvent this massive dimensionality problem and to account for the linkage disequilibrium (LD) structure between markers, Li *et al.* (2010) propose to restrict the analysis to a 1:3 case-pseudo-control matching based on LD blocks. For markers within a LD block, pseudo-controls are generated for the entire block using phased haplotypes, and three realizations of possible pseudo-controls are then randomly sampled (without replacement) from this collection. The resulting pseudo-control genotype blocks are then concatenated to complete the data frame for the analysis. These data can then be analyzed using logic regression with a conditional logistic link; that is, in each of the iterations of this logic regression algorithm, the parameters of the new model are estimated by the values that maximize the conditional logistic regression likelihood given by

$$\text{Likelihood}(\beta_1) = \prod_{t=1}^{n_{\text{Trio}}} \frac{\exp(\beta_1 \ell_{1t}^{(0)})}{\sum_{c=0}^{3} \exp(\beta_1 \ell_{1t}^{(c)})},$$

where ${\ell_{1t}}^{(0)}$ is the value of the logic expression L_1 for the case from the t-th trio, $t = 1, \ldots, n_{\text{Trio}}$, and ${\ell_{1t}}^{(c)}$ with $c = 1, 2, 3$ are the values of L_1 for the corresponding pseudo-controls. In the annealing procedure for the detection of good scoring models, the new model is compared with the current logic regression model using the negative log-likelihood as the objective function. To determine the appropriate model size, model selection is carried out using permutation tests (see Li *et al.*, 2010 for details). The software for trio logic regression has been implemented and will be available in the `R` packages `LogicReg`.

C. Monte Carlo logic regression

In logic regression, the search via simulated annealing leads to the identification of a single model. However, it is common that several good-scoring models exist that describe the association between the covariates and the outcome equally or almost equally well. This is in particular the case when some of the covariates are highly correlated, which is for example frequently the case in SNP association studies when multiple markers in a LD block are typed. Thus, many possible models of the same size might exist that have similar or even identical scores as assessed by the objective function. In these settings, it is often of interest to generate a list of covariates that are associated with the outcome when viewed as interacting with other covariates. For this purpose, Kooperberg and Ruczinski (2005) developed Monte Carlo logic regression, which employs a Metropolis-Hastings sampler based on reversible jump Markov chain Monte Carlo (Green, 1995). This procedure also identifies variables that most frequently occur jointly in the same model.

A geometric prior is used for the model size, but no priors are used for the actual model parameters. Instead, the maximum likelihood estimates for these parameters are computed. In each Monte Carlo logic regression iteration, a new model is proposed by randomly selecting one of the permissible moves in the move set (Fig. 2.3), and applying this move to one of the trees in the current logic regression model. The new model is accepted with a probability that depends on the prior ratio, the likelihood ratio, and the posterior ratio (see Green, 1995). The user selects a certain number of iterations for the "burn-in," and all iterations thereafter contribute to the statistical output. Recorded are how often a variable occurs in the sampled logic models, how often a variable pair jointly occurs in the sampled logic models, how often a trio of variables jointly occurs in the sampled logic models, etc. These variable occurrences provide the basis for a ranking similar to measures of variable importance. It is noteworthy that the stringency on the prior of model size can have an effect on how often any particular variable is included in the Boolean models seen in the sampler, however, it has also been observed that the actual variable rankings are typically not affected by the prior (Kooperberg and Ruczinski, 2005). Further statistics of interest include variable pair "enrichments," that is, the ratio of the number of jointly occurring variable pairs and their expected numbers under no interaction, derived from the marginal counts. Software for Monte Carlo logic regression is available as an option in the `R` function `logreg`, implemented in the `R` package `LogicReg`.

D. logicFS

Schwender and Ickstadt (2008) proposed a procedure called logicFS (logic Feature Selection) in which the original logic regression algorithm (see Algorithm 2) is applied to several subsets of a case-control data set. Similar to Monte Carlo logic regression (Kooperberg and Ruczinski, 2005), the objective is to identify different logic regression models and disease-associated SNP interactions. This approach, based on bagging (Breiman, 1996) with base learner logic regression, is described in Algorithm 3. The procedure stabilizes the search for disease-associated SNP interactions in a similar fashion as ensemble methods such as Random Forests (Breiman, 2001) stabilize and improve the prediction of classifiers such as CART.

Algorithm 3: Regression Approach of logicFS

Given: a data set with a binary response, and the number B of models that should be fitted.

1. For $b = 1, \ldots, B$,
   (a) draw a bootstrap sample from the subjects in the data set,
   (b) apply Algorithm 2 with g being the logit function to this bootstrap sample to fit a logic regression model,
   (c) convert each logic expression comprised by the logic regression model into a disjunctive normal form, that is, an OR-combination of AND-combinations, consisting of monomials, that is, AND-combinations, of minimal length,
   (d) compute the number N_b of out-of-bag observations, that is, observations that are not in the bootstrap sample, correctly classified by the fitted logic regression model.
2. For each detected monomial I_h, $h = 1, \ldots, n_{\mathrm{IA}}$,
   (e) remove I_h from all B logic regression models,
   (f) recompute the number $N_b^{(-h)}$ of out-of-bag observations correctly classified by the reduced model,

(g) determine the importance of the interaction I_h by

$$\mathrm{VIM}_{\mathrm{IA}}(I_h) = \frac{1}{B}\sum_{b=1}^{B}\left(N_b - N_b^{(-h)}\right) = \frac{1}{B}\sum_{b:I_h\in\Gamma_b}\left(N_b - N_b^{(-h)}\right),$$

where Γ_b is the set of all monomials/interactions comprised by the bth logic regression model.

The Boolean expressions are transformed into Disjunctive Normal Form to make the interactions in the corresponding logic regression models identifiable, as the monomials correspond to the SNP interactions. In logicFS, an importance measure similar to one of the variable importance measures (VIMs) in Random Forests is generated. The main difference is that in logicFS the interaction of interest is removed from the models to compute its importance, whereas Breiman (2001) permutes the values of the variable of interest once to determine the variable importance.

Further, extensions of logicFS for classification and for multicategorical and quantitative responses are available. These versions differ from Algorithm 3 only by the type of logic regression model fitted, however, slightly different importance measures for classification and for settings with a quantitative response have been devised (see Schwender and Ickstadt, 2008 for details). The logicFS algorithm has been implemented in the `R` package `logicFS`, and is freely available from Bioconductor, an open source and open development software project for the analysis and comprehension of genomic data (www.bioconductor.org).

E. Genetic programming for association studies

Genetic Programming for Association Studies (GPAS) proposed by Nunkesser *et al.* (2007) is applicable in classification settings, and uses genetic programming (Koza, 1993) as search algorithm. In contrast to logic regression, multivalued logic is used in GPAS. Thus, SNPs do not have to be coded as two binary variables, but can be used directly as input variables. Further, GPAS searches for logic expressions in Disjunctive Normal Form, which reduces the search space substantially.

Similar to logic regression, GPAS uses a tree representation of the Disjunctive Normal Forms (Fig. 2.4) and provides moves for alternating, removing, and adding literals to search for the logic expression that best predicts the binary outcome. In addition, GPAS provides moves for adding or removing monomials to or from the tree, and for combining two trees by adding a subtree of one of the trees to the other. Since genetic programming borrows terminology from Genetics, the GPAS moves are called "mutations" or "cross-over." The trees are referred to as individuals and each iteration of the algorithm is called a "generation."

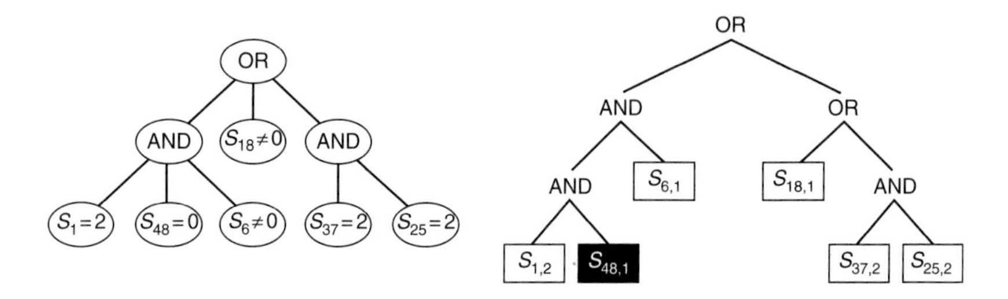

Figure 2.4. GPAS tree versus logic tree: the same classification rule based on six literals is displayed as a GPAS tree (left), and as a logic tree (right).

GPAS starts with a generation consisting of two individuals, each composed of a randomly selected literal. For each SNP S_i four literals are defined based on the number of minor alleles, namely $S_i = 0$, $S_i = 2$, $S_i \neq 0$, and $S_i \neq 2$. In the algorithm, all individuals from the current generation are used as candidates for the next generation. Each of the five available mutations is applied to a randomly selected individual and cross-over is applied to two randomly chosen individuals to combine those individuals. The resulting six individuals are also candidates for the next generation. The "fitness" according to an objective function is computed for each candidate, where the criteria of this objective function are the maximization of the mean proportion of correctly classified cases and controls, the maximization of the number of controls correctly predicted by the individual, and the minimization of the number of literals composing the individual. All individuals that are dominated by another individual (i.e., all individuals showing a lower value for each of these three criteria than one of the other individuals) are removed, and the remaining individuals form the next generation. These steps are repeated until either a predetermined number of generations has been reached, or the individuals do not change for a certain number of generations. Software for GPAS has been implemented in the `R` package `Rfreak`, which can be downloaded from the Comprehensive R Archive Network (http://cran.r-project.org/).

F. Finding haplotype combinations with an evolutionary algorithm

Clark *et al.* (2005, 2008) introduce a procedure that employs an evolutionary algorithm to search for models consisting of one or more Boolean expressions in which the literals are strings of SNPs in strong LD (e.g., indicating haplotype blocks). The Boolean expressions are represented as trees, similar to the trees in logic expression. While the leaves in each of these trees consist of strings of SNPs in strong LD, the SNPs in different leaves are in low LD or uncorrelated.

The move set of the multi-tree method is composed of mutations for modifying an operator, deleting one of the leaves, replacing a leaf by a string of SNPs currently not in the tree, shifting the string of SNPs, changing the number of SNPs in a leaf, and replacing the string of SNPs by another string. The move set also contains cross-over moves for adding a subtree of one tree to another tree, adding a new tree to the model, and combining leaves from two trees. These moves are randomly applied to the trees in each of the iterations, and the selection probabilities for the moves can differ between moves and generations.

After generating candidates for the next generation, a fitness function based on the marginal likelihood in a Bayesian regression framework is derived for each candidate model, and only the best models are selected for the next generation based on an "elitist" selection method (see Fogel and Corne, 2003). This procedure is repeated until convergence or a prespecified number of generations has been reached.

References

An, P., Feitosa, M., Ketkar, S., Adelman, A., Lin, S., Borecki, I., and Province, M. (2009). Epistatic interactions of CDKN2B-TCF7L2 for risk of type 2 diabetes and of CDKN2B-JAZF1 for triglyceride/high-density lipoprotein ratio longitudinal change: Evidence from the Framingham Heart Study. *BMC Proc.* **3**(Suppl. 7), S71.

Andrew, A. S., Karagas, M. R., Nelson, H. H., Guarrera, S., Polidoro, S., Gamberini, S., Sacerdote, C., Moore, J. H., Kelsey, K. T., Demidenko, E., *et al.* (2008). DNA repair polymorphisms modify bladder cancer risk: A multi-factor analytic strategy. *Hum. Hered.* **65,** 105–118.

Breiman, L. (1996). Bagging predictors. *Mach. Learn.* **26,** 123–140.

Breiman, L. (2001). Random forests. *Mach. Learn.* **45,** 5–32.

Breiman, L., Friedman, J. H., Olshen, R. A., and Stone, C. J. (1984). Classification and Regression Trees. Wadsworth, Belmont.

Carty, C. L., Heagerty, P., Heckbert, S. R., Jarvik, G. P., Lange, L. A., Cushman, M., Tracy, R. P., and Reiner, A. P. (2010). Interaction between fibrinogen and IL-6 genetic variants and associations with cardiovascular disease risk in the cardiovascular health study. *Ann. Hum. Genet.* **74,** 1–10.

Clark, T. G., De Iorio, M., Griffiths, R. C., and Farrall, M. (2005). Finding associations in dense genetic maps: A genetic algorithm approach. *Hum. Hered.* **60,** 97–108.

Clark, T. G., De Iorio, M., and Griffiths, R. C. (2007). Bayesian logistic regression using a perfect phylogeny. *Biostatistics* **8,** 32–52.

Clark, T. G., De Iorio, M., and Griffiths, R. C. (2008). An evolutionary algorithm to find associations in dense genetic maps. *IEEE Trans. Evol. Comp.* **12,** 297–306.

Cordell, H. J., Barratt, B. J., and Clayton, D. G. (2004). Case/pseudocontrol analysis in genetic association studies: A unified framework for detection of genotype and haplotype associations, gene-gene and gene-environment interactions, and parent-of-origin effects. *Genet. Epidemiol.* **26,** 167–185.

Culverhouse, R., Suarez, B. K., Lin, J., and Reich, T. (2002). A perspective on epistasis: Limits of models displaying no main effect. *Am. J. Hum. Genet.* **70,** 461–471.

Eliot, M., Azzoni, L., Firnhaber, C., Stevens, W., Glencross, D. K., Sanne, I., Montaner, L. J., and Foulkes, A. S. (2009). Tree-based methods for discovery of association between flow cytometry data and clinical endpoints. *Adv. Bioinformatics* **2009,** 235320.

Enquobahrie, D. A., Smith, N. L., Bis, J. C., Carty, C. L., Rice, K. M., Lumley, T., Hindorff, L. A., Lemaitre, R. N., Williams, M. A., Siscovick, D. S., *et al.* (2008). Cholesterol ester transfer protein, interleukin-8, peroxisome proliferator activator receptor alpha, and toll-like receptor 4 genetic variations and risk of incident nonfatal myocardial infarction and ischemic stroke. *Am. J. Cardiol.* **101,** 1683–1688.

Etzioni, R., Kooperberg, C., Pepe, M., Smith, R., and Gann, P. H. (2003). Combining biomarkers to detect disease with application to prostate cancer. *Biostatistics* **4,** 523–538.

Etzioni, R., Falcon, S., Gann, P. H., Kooperberg, C. L., Penson, D. F., and Stampfer, M. J. (2004). Prostate-specific antigen and free prostate-specific antigen in the early detection of prostate cancer: Do combination tests improve detection? *Cancer Epidemiol. Biomarkers Prev.* **13,** 1640–1645.

Feng, Q., Balasubramanian, A., Hawes, S. E., Toure, P., Sow, P. S., Dem, A., Dembele, B., Critchlow, C. W., Xi, L., Lu, H., *et al.* (2005). Detection of hypermethylated genes in women with and without cervical neoplasia. *J. Natl. Cancer Inst.* **97,** 273–282.

Fogel, G. B., and Corne, D. W. (2003). An introduction to evolutionary computation for biologists. *In* "Evolutionary Computation in Bioinformatics" (G. B. Fogel and D. W. Corne, eds.), pp. 19–38. Morgan Kaufmann, San Francisco.

Fritsch, A., and Ickstadt, K. (2007). Comparing logic regression based methods for identifying SNP interactions. *In* "Bioinformatics in Research and Development" (S. Hochreiter and R. Wagner, eds.), pp. 90–103. Springer, Berlin.

Garte, S. (2001). Metabolic susceptibility genes as cancer risk factors: Time for a reassessment? *Cancer Epidemiol. Biomarkers Prev.* **10,** 1233–1237.

Green, P. J. (1995). Reversible jump Markov chain Monte Carlo computation and Bayesian model determination. *Biometrika* **82,** 711–732.

Harth, V., Schaefer, M., Abel, J., Maintz, L., Neuhaus, T., Besuden, M., Primke, R., Wilkesmann, A., Thier, R., Vetter, H., *et al.* (2008). Head and neck squamous-cell cancer and its association with polymorphic enzymes of xenobiotic metabolism and repair. *J. Toxicol. Environ. Health A* **71,** 887–897.

Hosmer, D. W., and Lemeshow, S. (1989). Applied Logistic Regression. Springer, New York.

Huang, J., Lin, A., Narasimhan, B., Quertermous, T., Hsiung, C. A., Ho, L. T., Grove, J. S., Olivier, M., Ranade, K., Risch, N. J., and Olshen, R. A. (2004). Tree-structured supervised learning and the genetics of hypertension. *Proc. Natl. Acad. Sci. USA* **101,** 10529–10534.

Ickstadt, K., Schaefer, M., Fritsch, A., Schwender, H., Abel, J., Bolt, H. M., Bruening, T., Ko, Y. D., Vetter, H., and Harth, V. (2008). Statistical methods for detecting genetic interactions: A head and neck squamous-cell cancer study. *J. Toxicol. Environ. Health A* **71,** 803–815.

Janes, H., Pepe, M., Kooperberg, C., and Newcomb, P. (2005). Identifying target populations for screening or not screening using logic regression. *Stat. Med.* **24,** 1321–1338.

Justenhoven, C., Hamann, U., Schubert, F., Zapatka, M., Pierl, C. B., Rabstein, S., Selinski, S., Mueller, T., Ickstadt, K., Gilbert, M., *et al.* (2008). Breast cancer: A candidate gene approach across the estrogen metabolic pathway. *Breast Cancer Res. Treat.* **108,** 137–149.

Keles, S., van der Laan, M. J., and Vulpe, C. (2004). Regulatory motif finding by logic regression. *Bioinformatics* **20,** 2799–2811.

Kirkpatrick, S., Gelatt, C. D. J., and Vecchi, M. P. (1983). Optimization by simulated annealing. *Science* **220,** 671–680.

Kooperberg, C., and Ruczinski, I. (2005). Identifying interacting SNPs using Monte Carlo logic regression. *Genet. Epidemiol.* **28,** 157–170.

Kooperberg, C., Ruczinski, I., LeBlanc, M., and Hsu, L. (2001). Sequence analysis using logic regression. *Genet. Epidemiol.* **21,** 626–631.

Kooperberg, C., Bis, J. C., Marciante, K. D., Heckbert, S. R., Lumley, T., and Psaty, B. M. (2007). Logic regression for analysis of the association between genetic variation in the renin–angiotensin system and myocardial infarction or stroke. *Am. J. Epidemiol.* **165,** 334–343.

Koza, J. R. (1993). Genetic Programming—On the Programming of Computers by Means of Natural Selection. MIT Press, Cambridge.

Li, Q., Fallin, M. D., Louis, T. A., Lasseter, V. K., McGrath, J. A., Avramopoulos, D., Wolyniec, P. S., Valle, D., Liang, K. Y., Pulver, A. E., and Ruczinski, I. (2010). Detection of SNP–SNP interactions in trios of parents with schizophrenic children. *Genet. Epidemiol.* **34**(5), 396–406.

Marchini, J., Donnely, P., and Cardon, R. C. (2005). Genome-wide strategies for detecting multiple loci that influence complex diseases. *Nat. Genet.* **37,** 413–416.

Nicodemus, K. K., Callicott, J. H., Higier, R. G., Luna, A., Nixon, D. C., Lipska, B. K., Vakkalanka, R., Giegling, I., Rujescu, D., St. Clair, D., *et al.* (2010). Evidence of statistical epistasis between DISC1, CIT and NDEL1 impacting risk for schizophrenia: Biological validation with functional neuroimaging. *Hum. Genet.* **127,** 441–452.

Nunkesser, R., Bernholt, T., Schwender, H., Ickstadt, K., and Wegener, I. (2007). Detecting high-order interactions of single nucleotide polymorphisms using genetic programming. *Bioinformatics* **23,** 3280–3288.

Poole, E. M., Hsu, L., Xiao, L., Kulmacz, R. J., Carlson, C. S., Rabinovitch, P. S., Makar, K. W., Potter, J. D., and Ulrich, C. M. (2010). Genetic variation in prostaglandin e2 synthesis and signaling, prostaglandin dehydrogenase, and the risk of colorectal adenoma. *Cancer Epidemiol. Biomarkers Prev.* **19,** 547–557.

R Development Core Team (2009). R: A Language and Environment for Statistical Computing. R Foundation for Statistical Computing, Vienna, Austria 3-900051-07-0.

Ruczinski, I. (2000). Logic Regression and Statistical Issues Related to the Protein Folding Problem. Ph.D. thesis. Department of Statistics, University of Washington, Seattle.

Ruczinski, I., Kooperberg, C., and LeBlanc, M. (2003). Logic regression. *J. Comput. Graph. Stat.* **12,** 475–511.

Ruczinski, I., Kooperberg, C., and LeBlanc, M. (2004). Exploring interactions in high-dimensional genomic data: An overview of logic regression, with applications. *J. Mult. Anal.* **90,** 178–195.

Schaid, D. J. (1996). General score tests for associations of genetic markers with disease using cases and their parents. *Genet. Epidemiol.* **13,** 423–449.

Schaid, D. J. (1999). Likelihoods and TDT for the case-parents design. *Genet. Epidemiol.* **16,** 250–260.

Schwender, H., and Ickstadt, K. (2008). Identification of SNP interactions using logic regression. *Biostatistics* **9,** 187–198.

Segal, M. R., Barbour, J. D., and Grant, R. M. (2004). Relating HIV-1 sequence variation to replication capacity via trees and forests. *Stat. Appl. Genet. Mol. Biol.* **3,** 2.

Suehiro, Y., Wong, C. W., Chirieac, L. R., Kondo, Y., Shen, L., Webb, C. R., Chan, Y. W., Chan, A. S. Y., Chan, T. L., Wu, T. T., *et al.* (2008). Epigenetic–genetic interactions in the apc/wnt, ras/raf, and p53 pathways in colorectal carcinoma. *Clin. Cancer Res.* **14,** 2560–2569.

Vaidya, V. S., Waikar, S. S., Ferguson, M. A., Collings, F. B., Sunderland, K., Gioules, C., Bradwin, G., Matsouaka, R., Betensky, R., Curhan, G. C., and Bonventre, J. V. (2008). Urinary biomarkers for sensitive and specific detection of acute kidney injury in humans. *Clin. Transl. Sci.* **3,** 200–208.

Witte, J. S., and Fijal, B. A. (2001). Introduction: Analysis of sequence data and population structure. *Genet. Epidemiol.* **21,** 600–601.

Yaziji, H., Battifora, H., Barry, T. S., Hwang, H. C., Bacchi, C. E., McIntosh, M. W., Kussick, S. J., and Gown, A. M. (2006). Evaluation of 12 antibodies for distinguishing epithelioid mesothelioma from adenocarcinoma: Identification of a three-antibody immunohistochemical panel with maximal sensitivity and specificity. *Mod. Pathol.* **19,** 514–523.

3

Complex System Approaches to Genetic Analysis: Bayesian Approaches

Melanie A. Wilson, James W. Baurley, Duncan C. Thomas, and David V. Conti

Department of Preventive Medicine, University of Southern California, Los Angeles, California, USA

ABSTRACT

Genetic epidemiology is increasingly focused on complex diseases involving multiple genes and environmental factors, often interacting in complex ways. Although standard frequentist methods still have a role in hypothesis generation and testing for discovery of novel main effects and interactions, Bayesian methods are particularly well suited to modeling the relationships in an integrated "systems biology" manner. In this chapter, we provide an overview of the principles of Bayesian analysis and their advantages in this context and describe

Advances in Genetics, Vol. 72

0065-2660/10 $35.00
DOI: 10.1016/S0065-2660(10)72003-4

various approaches to applying them for both model building and discovery in a genome-wide setting. In particular, we highlight the ability of Bayesian methods to construct complex probability models via a hierarchical structure and to account for uncertainty in model specification by averaging over large spaces of alternative models.

I. INTRODUCTION

Bayesian approaches have gained a tremendous amount of popularity in a diverse set of applications including but not limited to: economics, environmental science, bioinformatics, epidemiology, genetics, computer science, political science, and public policy. Within these fields, the Bayesian framework can be applied to a wide range of statistical model classes such as linear regression, generalized linear models, survival analysis, tree models, graphical models, and spatial analyses. The growth in popularity of Bayesian approaches is due in most part to the intuitive nature of inference within the framework, the extreme flexibility of the models, and the computational developments helping to facilitate practical analyses. While this chapter focuses more specifically on the use of Bayesian approaches for complex genetics applications, we begin with a general introduction to the fundamentals of any Bayesian analysis.

A. Fundamentals of a Bayesian approach

The fundamentals of a Bayesian approach lie in Bayes Rule, which is the tool that allows us to revise our current set of beliefs about unknown parameters given a set of observed data Y via conditional probabilities:

$$p(\theta|Y) = \frac{p(\theta, Y)}{p(Y)} = \frac{p(Y|\theta)p(\theta)}{\int p(Y|\theta)p(\theta)d\theta}$$

where the integral in the denominator can be replaced by a summation if the probability distribution of θ is discrete. Thus, any Bayesian approach has two major components: (1) defining the joint probability model $p(\theta, Y)$ and (2) computing conditional probabilities $p(\theta \mid Y)$. In defining the joint probability model, we must specify the likelihood of the observed data given the parameters of interest, $p(Y \mid \theta)$. This specification is common to both a frequentist approach and a Bayesian approach. However, instead of assuming that the parameters of the model are fixed and their true value is unknown, the Bayesian framework assumes that the parameters themselves are random variables. Thus, to define the joint probability model, we must also specify the prior distribution of the parameters, $p(\theta)$, in addition to the likelihood of the observed data.

The above framework assumes that we are interested in making inference on all of the parameters in the probability model. However, in many applications this is not the case. If there is some subset of parameters, θ_I, that we are interested in and the remaining, θ_N, are nuisance parameters we can rewrite the conditional probability for the parameters of interest as:

$$p(\theta_I|Y) = \frac{p(Y|\theta_I)p(\theta_I)}{p(Y)},$$

where $p(Y \mid \theta_I)$ is the marginal likelihood of the data given the parameters of interest for making inference and can be calculated by integrating out the nuisance parameters (or summation for discrete measures):

$$\int p(Y|\theta_I, \theta_N)p(\theta_N)d\theta_N.$$

Thus, computing conditional probabilities in the Bayesian framework often requires computing high-dimensional integrals (or summations) for both the marginal likelihoods for the parameters of interest, $p(Y \mid \theta_I)$, and the normalizing constant, $p(Y)$.

Finally, in many analyses, there may exist several alternative specifications for the joint model. In a Bayesian analysis, we can incorporate the uncertainty of the specification of the joint model by indexing a specific model by an indicator vector γ. Since we are now considering multiple models, $M_\gamma \in M$, we can rewrite the joint probability statement as:

$$p(Y, \theta_\gamma, M_\gamma) = p(Y|\theta_\gamma, M_\gamma)p(\theta_\gamma|M_\gamma)p(M_\gamma)$$

where θ_γ are the parameters specific to model M_γ. We can therefore make inference on the models themselves by calculating the conditional probabilities:

$$p(M_\gamma|Y) = \frac{p(Y|M_\gamma)p(M_\gamma)}{\sum_{M_\gamma \in M} p(Y|M_\gamma)p(M_\gamma)}$$

where $p(Y \mid M_\gamma) = \int p(Y \mid \theta_\gamma, M_\gamma)p(\theta_\gamma)d\theta_\gamma$ is the marginal likelihood of the observed data given any of the models of interest.

B. Bayesian advantages

Bayesian approaches come with many advantages. First and foremost, by specifying a probability distribution on the parameters we directly quantify the uncertainty in those parameters given the observed data and achieve statistical conclusions with common sense interpretations. These probability statements allow for a conceptually straightforward approach to inference, in that our prior beliefs specified in $p(\theta)$ are updated based on the observed data, Y, via conditional probabilities $p(\theta \mid Y)$. This is quite different from a frequentist approach. A simple example is the Bayesian credible interval versus the frequentist confidence interval. A frequentist 95% confidence interval must be interpreted based on hypothetical repeated sampling in which S repeated samples are taken from the population and, subsequently, $S \times 0.95$ of the estimated parameters would fall within the confidence interval. In contrast, the Bayesian 95% credible interval is interpreted potentially more intuitively as a 95% probability that the true value of the parameter lies within the calculated credible interval.

Another advantage of the Bayesian framework is that it provides a very natural setting for incorporating complex structures, multiple parameters, and procedures for dealing with nuisance parameters (parameters that we do not wish to make inference about). The only restriction within a Bayesian approach is that one must be able to specify a joint probability model for the observed data and parameters of interest. We can therefore include as many parameters to our models as needed and simply marginalize across (i.e., integrate out or sum over) the ones that we are not interested in making inference on. We are also able to incorporate external information in the analysis in an explicit manner by specifying prior probability distributions for the parameters of interest. This is particularly useful in the biological setting where there is often a great deal of external information and incorporating this information can potentially help the practitioner narrow the focus of an otherwise overly complex model. Finally, the Bayesian framework provides a natural setting for incorporating model uncertainty into any analysis by extending the hierarchy and viewing the model itself as a random variable with its own prior distribution.

C. Limitations

Many of the main advantages of a Bayesian approach lie in the specification of prior distributions on the model parameters and in some cases on the models themselves. However, this can also be one of the main limitations of the Bayesian framework. The prior distributions can be specified in either a subjective or objective manner depending about the amount of prior knowledge or external information one has for the parameters or the acceptable degree to which the posterior results are sensitive to the prior specification. However, even if there is a large amount of prior information regarding the parameters of interest or the models themselves, it is not always straightforward to quantify a practitioner's prior knowledge and elicit prior distributions. Also, if a limited amount of external knowledge exists about the parameters, the question remains about how to specify the priors in an objective ("noninformative") manner (Thomas *et al.*, 2007a). Advances have been made in both the elicitation of prior distributions and in developing and investigating the asymptotic characteristics of objective priors on both the parameters of interest and the models themselves (discussed in more detail in Sections III and IV).

Another limitation to the Bayesian framework is the complexity of computing conditional probabilities. This complexity lies in the difficulty in performing potentially high-dimensional integrals (or summations) in both the normalizing constant $p(Y)$ and in the marginal likelihood of the data $p(Y \mid \theta_I)$ or $p(Y \mid M_\gamma)$. Because of the computational constraints on Bayesian approaches, practitioners were historically limited to choosing only conjugate likelihoods and priors for which conditional probabilities could be calculated in closed form. However, with recent advances in estimating high-dimensional integrals with stable numerical approximations or by simulation via Markov Chain Monte Carlo (MCMC), Metropolis–Hastings (MH), and stochastic model search algorithms, one is much less limited when choosing a joint probability model. These technical advances, as well as software to more easily implement Bayesian models (such as WinBUGS (Speigelhalter *et al.*, 2003) and JAGS (http://sourceforge.net/projects/mcmc-jags)), have lead to an increased popularity of Bayesian approaches (discussed in more detail in Sections III and IV).

D. General model specification

The remainder of the chapter utilizes a generalized linear model framework in which most any type of outcome variable may be analyzed with an appropriate link function. However, for simplicity, we will focus on a binary outcome variable. Let Y be a vector of length n comprised of some binary outcome variable for individual i with expected value

$$E(Y) = \mu$$

with mean vector $\mu = (\mu_1, \cdots, \mu_n)^T$. Also, for each individual, we assume that p covariates are measured, $x_1, \cdots, x_p$. We then use generalized regression models to relate the binary outcome variable to a subset of predictor variables. We denote the collection of all possible models by M. An individual model, denoted by M_γ, is specified by an indicator vector γ. Then under each model M_γ, μ is of the general form:

$$M_\gamma : \mu = g^{-1}\left[f(X_\gamma)\beta_\gamma\right]$$

where g is the link function (usually logit whenever Y is binary), $f(X_\gamma)$ is some general structure or parameterization of a set of covariates, and β_γ are the effects of $f(X_\gamma)$ on the outcome of interest Y (are parameters of interest θ_γ). A very general structure for the models can be seen as a directed graph in Fig. 3.1. Specifically in this example, the γ indicator allows both the covariates x_1 and x_3 to inform the model and design matrix X_γ, whereas all other covariates (e.g., x_2 and x_p) are excluded. Then $f(X_\gamma)$ defines the structure or parameterization of these variables in the regression. Finally, the link function, $g^{-1}(\cdot)$, as well as the regression coefficients, β_γ, relate the structure built in $f(X_\gamma)$ to disease.

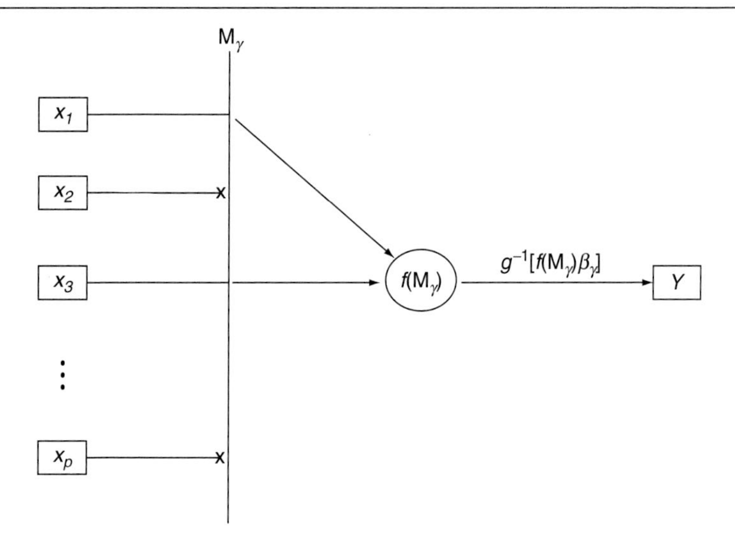

Figure 3.1. Directed graph summarizing general model structure.

For a simple illustration of the general model structure, we consider a simple logistic regression analysis where the link function is defined as:

$$\mu = \frac{\exp\{f(X_\gamma)\beta_\gamma\}}{1 + \exp\{f(X_\gamma)\beta_\gamma\}}$$

Each model is defined by the indicator γ denoting which of the p possible covariates will be included in the marginal model. In particular, we can have γ be an indicator vector with $\gamma_j = 1$ if covariate j is the one included in the marginal model and $\gamma_i = 0$ for every $i \neq j$. Then the model space M is made up of p single variable models M_γ. Under each model $M_\gamma \in M$ the mean vector has the form:

$$M_\gamma : \ln\left(\frac{\mu}{1-\mu}\right) = \beta_0 + X_\gamma \beta_\gamma$$

where X_γ is of dimension $n \times 1$ and β_γ is the logarithm of the odds ratio for the single covariate in model M_γ. In the simple marginal case, it is plausible to perform an enumeration of the entire model space M, so typically no stochastic model search algorithm is needed.

Given the general structure that we have developed above, the models that we describe herein can be much more complex than this simple marginal example. In Section II, we review strategies for defining the structure or parameterization of the covariates that is specified in $f(X_\gamma)$. These strategies can range from the simple marginal analyses to the more complex networks and mechanistic models. In Section III, we tackle the problem of (1) what priors should be placed on the parameters of interest, β_γ, and (2) how to compute the conditional probabilities and estimate β_γ. In Section IV, we incorporate an additional layer into the hierarchy by investigating multiple models $M_\gamma \in M$. Section IV mainly focuses on (1) defining prior distributions on the model space $p(M_\gamma)$, (2) how to approximate high-dimensional integrals for marginal likelihoods for the models of interest, $p(Y \mid M_\gamma)$, and (3) approximating high-dimensional sums for normalizing constants. Finally, in Section V, we describe methods for determining noteworthiness of the structures defined in Section II and marginally of the covariates themselves.

II. STRUCTURE OF COVARIATES

We begin with the simplest case of a single polymorphism in a single gene, with no genetic or environmental modifiers. In this case, X is comprised of genotypes at a single polymorphism for n individuals and the only consideration in specifying the covariate function $f(X)$ concerns issues of dominance. If the locus is diallelic, with alleles denoted a and A, then we might wish to consider any of the following codings for the genotype X_i:

$f(X_i) = 1 \text{ if } X = AA \text{ or } Aa, 0 \text{ if } X = aa$	Dominant
$f(X_i) = 2 \text{ if } X = AA, 1 \text{ if } X = Aa, 0 \text{ if } X = aa$	Additive
$f(X_i) = 1 \text{ if } X = AA, 0 \text{ if } X = Aa \text{ or } aa$	Recessive
$f(X_i) = 1 \text{ if } X = AA, \theta \text{ if } X = Aa, 0 \text{ if } X = aa$	Codominant

(Note that the codominant model entails an additional parameter θ that would be estimated along with the other regression coefficients in β. This is most easily accomplished by fitting a 2-degree of freedom model with two dummy variables for each genotype). In some instances, the choice of coding might be determined by prior biological knowledge, but in the absence of relevant knowledge one would consider each of these possibilities within the general model selection or model uncertainty framework developed in greater detail in Section III.

Most genes have numerous polymorphisms and any candidate pathway study might consider a broad range of genes, so multivariate models are called for. An obvious choice for a binary disease trait might be a logistic regression model

$$\text{logit}(p(Y = 1|X)) = \beta_0 + \sum \beta_j f_j(X_j)$$

Here, we could incorporate different genetic codings for each variable and therefore we allow the function $f_j(X_j)$ to differ across the covariates. It might also be helpful to consider a hierarchical structure for multiple polymorphisms within genes within pathways. Letting $j = 1, \ldots, J$ index the pathways, and $k = 1, \ldots, K_j$ index the genes within pathway j, and $l = 1, \ldots, L_{jk}$ polymorphisms within genes, one might extend the logistic model as

$$\text{logit}(p(Y = 1|X)) = \beta_0 + \sum_j \beta_j \sum_k \beta_{j_k} \sum_l \beta_{j_{k_l}} f_{j_{k_l}}\left(X_{j_{k_l}}\right)$$

with higher level models for the βs constructed in a manner to ensure identifiability. Again, working within the general modeling framework discussed below, one might wish to consider different strategies for selection, shrinkage, or averaging at the polymorphism, gene, or pathway levels.

Gene–environment (G×E) and gene–gene (G×G) interactions introduce a further level of complexity, but the same general framework can be applied. Consider, for example, a model for the joint effect of two polymorphisms X_1 and X_2. The logistic model could then be written as

$$\text{logit}(p(y = 1|X)) = \beta_0 + \beta_1 f_1(X_1) + \beta_2 f_2(X_2) + \beta_{12} f_{12}(X_{12})$$

where $f_1(X_1)$ and $f_2(X_2)$ denote any of the genetic codings discussed above and $f_{12}(X_{12})$ could be a simple product X_1X_2 or a term constructed to capture phase information for two polymorphisms in linkage disequilibrium (Conti and Gauderman, 2004), or it could be some more complex epistasis model, such as 1 for X_{12}=(*aaBB*, *aAbB*, *AAbb*), 0 otherwise. See Li and Reich (2000), Moore (2003), Moore *et al.* (2007), Zhang and Liu (2007), Cordell (2009), Moore and Williams (2009), and Tang *et al.* (2009) for further discussion of various possibilities for modeling epistasis. Similar codings for G×E interactions are possible (Thomas, 2010a,b). For multiple polymorphisms within the same gene, one might use a haplotype-based model, $\text{logit}(p(Y=1 \mid X)) = \beta_0 + \beta_1 h_1(X) + \beta_2 h_2(X)$, where h_1 and h_2 represent the two haplotypes carried by an individual comprising the set of alleles at the different loci carried on the same chromosome. In the absence of phase information, this would require forming a likelihood by summing over all possible arrangements of the alleles into haplotypes, weighted by their probabilities based on population linkage disequilibrium patterns (Stram *et al.*, 2003).

If specific prior knowledge of the relevant biological process is known, mechanistic approaches may be used to construct more complex models. These models will generally be quite specific to a particular pathway and their mathematical form will depend upon the nature of the pathway. Metabolic pathways, for example, might be modeled in terms of a series of latent variables $L(X;m)$ representing concentrations of intermediate metabolite m. These concentrations could be given by a system of differential equations based on the known pharmacokinetics with rates that depend in some manner on the genotypes encoding the relevant enzymes and their environmental substrates or cofactors (Thomas *et al.*, 2010). The solution to this system then provides a mathematical expression for the covariate function $f(X,\theta)$ where X now represents all the genetic and environmental inputs to the system and θ a vector of additional parameters to be estimated for a particular model. For example, Cortessis and Thomas (2003) described a metabolic model for the metabolism of polycyclic aromatic hydrocarbons and heterocyclic amines derived from tobacco smoke and well done red meat as risk factors for colorectal polyps. Each pathway involves several intermediate steps, metabolized by several genes. Using linear kinetics, the metabolic rate parameters were assumed to be lognormally distributed around population means specific to the relevant genotypes. The expected concentration of the final metabolites from each pathway was computed specific to each person's exposure and genotype, and used as covariates in a logistic model for disease. The model was fitted using MCMC methods, sampling individual metabolic rates and model parameters (regression coefficients, genotype-specific mean rates, and variances) in turn. Further elaboration of this model and comparisons with a BMA analysis using hierarchical models are provided in Conti *et al.* (2003). As before, one might wish to consider a range of alternative models, such as submodels of a general model including only some subset of the inputs or different codings of dominance or even different mathematical models (e.g., linear vs. Michaelis–Menten kinetics). These approaches have been most extensively developed in the context of population pharmacokinetic models (Best *et al.*, 1995; Bois, 2001; Clewell *et al.*, 2002; Davidian and Gallant, 1992; Gelman *et al.*, 1996; Lunn *et al.*, 2009; Racine-Poon and Wakefield, 1998; Wakefield, 1996), although so far, there has been relatively little attention to genetic variation in metabolic parameters.

III. ESTIMATION

So far, in this chapter, we have focused on how to link and combined the observed data into a probability model. However, the ability to measure numerous factors present many difficulties for analysis (Greenland, 1993). Conventional approaches have relied on either fitting a full model with all the factors included in the probability model or fitting a reduced model determined by some eliminating algorithm. Including all the factors in one model can lead to biased and unreliable estimates due to sparse data when the number of parameters approaches the number of individuals in the sample (Greenland, 2000a,b). Reduced models may avoid this complication, but they fail to account for the correlations that exist between all the factors and can lead to underestimated variance (Robins and Greenland, 1986). Additionally, when statistical tests are used for the numerous exposures, issues of multiple comparisons arise (Thomas *et al.*, 1985). However the model γ is determined, there will be corresponding parameters β_γ that describe the effect on the outcome of interest. For this part of the chapter, we focus on how prior specification on these parameters via hierarchical models can be used to construct complex models for analysis. For example, in a genome-wide association study (GWAS), one might have little or no external information about most SNPs, so such analyses are commonly treated as exploratory; indeed, their "agnostic" or "hypothesis free" nature is commonly touted as one of their advantages. Here, treating each polymorphism as independent may be appropriate. A pathway-driven study, on the other hand, may be able to exploit extensive knowledge about the pharmacokinetics or pharmacodynamics of the pathway. Recently, there has been an intriguing convergence of the two philosophies, with external pathway information being exploited to mine GWAS data for gene sets whose components may not separately attain genome-wide levels of significance but combination implicates particular pathways (Wang *et al.*, 2007; Zhong *et al.*, 2010), or by exploring GWAS data to discover hitherto unsuspected sets of genes that may share a common biological function (Sebastiani *et al.*, 2005).

Rather than treating each factor independently, specification of the relations among the observed data with two or more stages can be used to create an intricate joint probability model. While each stage may be relatively simple and easy to understand, the entire model may be much more sophisticated, with the aim to more accurately model the underlying complex processes. Additionally, by providing a joint probability model for all exposures, hierarchical modeling offers a potential solution to problems of multiple comparisons (Greenland and Robins, 1991; Thomas *et al.*, 1985). As one of the first examples in genetic epidemiology, Thomas *et al.* (1992) used hierarchical modeling to jointly evaluate numerous human leukocyte antigen (HLA) alleles and their association to insulin-dependent diabetes mellitus (IDDM), while incorporating several environmental risk factors.

The addition of higher stage information can substantially improve the accuracy and stability of effect estimates (Greenland, 2000a,b; Greenland and Poole, 1994; Morris, 1983). However, to achieve this improvement, the model hierarchy must be specified in a manner that efficiently uses the data and is scientifically plausible (Rothman and Greenland, 1998).

While a conventional model fits a single probability level to describe the relation between the multiple factors and the outcome, hierarchical models incorporate higher level prior distributions to explain the relations among the parameters. Although multiple levels of prior structure can be constructed, in practice there is a limit to the number of levels that can be feasibly estimated from the data without relying too much on a strongly specified prior. Consider the general scenario with multiple factors, X, and outcome, Y. The first-probability model at the individual-level can be specified as:

$$Y|X, \beta \sim p(Y|X, \beta)$$

As before, Y and X are observed data and β are the corresponding coefficients of risk associated with a one-unit increase for each factor. Subsequently, probability models are placed on the parameters, β, assuming that they come from a common probability distribution.

$$\beta|\theta \sim p(\beta|\theta)$$
$$\theta \sim p(\theta)$$

Such a hierarchy can apply dependencies to the estimates of the parameters β through the structure of $p(\beta \mid \theta)$. These dependencies are often modeled conditionally on certain parameters, θ, called hyperparameters and these hyperparameters can also be assigned probability distributions, $p(\theta)$. An important assumption here is that the parameters are exchangeable. That is, the parameters $(\beta_1,\ldots,\beta_m)$ are exchangeable in their joint distribution if $p(\beta_1,\ldots,\beta_m)$ is invariant to permutations of the indexes $(1,\ldots,m)$ (Gelman, 1995). Consequently, given no other information, other than the data itself, there is no prior ordering or grouping of the parameters $(\beta_1,\ldots,\beta_m)$. If this assumption holds, we may assume that the corresponding coefficients for the exposures, β, are drawn from the same common population distribution.

The general hierarchical modeling framework presented in the above equations provides a quite flexible framework to model complex systems. For example, $p(\beta \mid \theta)$ specified as a normal distribution centered at zero is akin to ridge regression, a frequentist robust regression technique (Sorenson and Gianola, 2002). Similarly, a double exponential for $p(\beta \mid \theta)$ is equivalent to the Lasso procedure (Park and Casella, 2008; Chen *et al.*, 2010). To demonstrate the flexibility and nuances of specifying such a hierarchy, consider a specific analysis of case-control data with numerous genetic and environmental factors. A specific hierarchical model can be specified as:

$$\text{logit}(p(Y = 1|X)) = \beta_0 + \sum \beta_j f_j(X_j)$$
$$\beta \sim \text{MVN}(\mathbf{Z}\pi, \Sigma)$$

where X is a design matrix of genetic and environmental factors for the individuals within the study, Z contains second-stage covariates for each of the environmental and genetic factors reflecting higher level, often prespecified relations, π is a column vector of coefficients corresponding to these higher level effects on disease, and Σ is a matrix specifying the residual covariance of the second-stage covariates. There are many possible types of information that could be used for defining Z. In genetics, this could define simple indicator variables for which pathway(s) each gene is thought to be active (Hung *et al.*, 2004), information extracted systematically from genomic or pathway ontologies (Conti *et al.*, 2009; Thomas *et al.*, 2007b), experimental information such as eQTLs from cell cultures (Zhong *et al.*, 2010), *in silico* predictions based on evolutionary conservation or predicted effect on protein conformation (Rebbeck *et al.*, 2004), or predictions from simulations of the pathway (Thomas *et al.*, 2010). Of course, multiple sources could, in principle, be combined and additional levels can easily be incorporated, such as separate models for SNPs within genes and genes within pathways (Conti and Gauderman, 2004) or for main effects and interactions (Conti *et al.*, 2003).

Overall this hierarchy will result in posterior estimates for the association parameters that are an inverse-variance weighted average between the conventional estimates from the logistic regression only and the estimated conditional second-stage means, $Z\pi$. To see this more clearly, consider a simple weighted regression approach for estimation. Here, the second-stage estimated prior means, $Z\tilde{\pi}$, and corresponding estimated covariance matrix, $(Z^{T}WZ)^{-1}$, can be obtained from a weighted least squares regression, $\tilde{\pi} = \left(Z^{T}WZ\right)^{-1}ZW\hat{\beta}$, where $W=(V+\Sigma)^{-1}$, and V is a diagonal matrix with elements equal to the square of the estimated standard errors for $\hat{\beta}$ (Morris, 1983). Averaging the first- and second-stage estimates yields posterior estimates

$$\tilde{\beta} = BZ\tilde{\pi} + (I - B)\hat{\beta}.$$

The estimate of the covariance matrix is given by $\tilde{C} = \hat{V}(I - (I - H)^{T}B)$, where $H=Z(Z^{T}WZ)^{-1}Z^{T}W$ and B is the estimated shrinkage matrix

$$B = WV = (V + \Sigma)^{-1}V$$

From these equations, we can see that if the maximum likelihood first-stage estimates, $\hat{\beta}$, have large variance, $\hat{V}$, relative to the prior variance, Σ, then B will also be large. As a result, the hierarchical analysis has several important differences from the conventional, single-stage logistic regression analysis. First, the conventional analysis has no constraints on the first-stage regression coefficients, β. For example, if we model the effects of smoking on lung cancer, we would expect the effect estimate, $\hat{\beta}$, to be quite high. But even in this extreme case, we would not expect outrageously high odds ratios. It makes sense to incorporate this expectation into our analysis by including a probability distribution for the effect estimates that weights more probable estimates higher. This may be done by setting the elements in the Z matrix to zero and allowing Σ to reflect our prior beliefs about the extent of the probability distribution of the first-stage coefficients, β. However, if we have information regarding the relations between the factors, we may incorporate that information into the Z design matrix and account for the effects due to these dependencies. In this case, Σ reflects residual covariance or associations between the first-stage parameters, β, after accounting for the relations defined in Z. If we allow the elements of Σ to go to infinity—that is, we have no prior belief regarding the distribution of the residual effects—the final estimates of effect will disregard any second-stage information and be equal to the first-stage conventional maximum likelihood estimates. If we set the elements of Σ to zero, believing that there are no residual effects beyond the relationships defined in Z, then the final estimates will be equal to the estimated second-stage conditional means, $Z\pi$. Thus, Σ acts as a smoothing parameter, controlling the amount of shrinkage from the conventional first-stage maximum likelihood estimates toward the second-stage conditional means. Within this context, the specification of Σ is very flexible. Joint modeling is often represented by setting $\Sigma = \tau^2 D$, where $\tau^2 > 0$ represents the overall variance of the effect estimates and **D** is a positive definite correlation matrix describing the correlation between the effect estimates for any two pairs of polymorphisms. Conti and Witte (2003) utilized this specification to set **D** to the expected decay of linkage equilibrium. Closely related to this is a Gaussian conditional autoregression specification where $D = (I - \rho A)^{-1}$ (Wakefield *et al.*, 2000). Here, the elements within **A** describe the spatial weights between any two polymorphisms and ρ is interpreted as the strength of the spatial dependence. Thomas *et al.* (2010) recently investigated the use of this model in the genetics context using prior gene–gene connection information such as from gene coexpression experiments to define **A**. The normal *g*-prior defines the covariance matrix as the inverse of the Fisher information matrix multiplied by a constant *g*, $\Sigma = g(X^T X)^{-1}$ (Zellner, 1986). This ensures that the correlation structure in the prior distribution matches the structure in the likelihood with either *g* being determined via information criterion (such as AIC) or estimated with a prior specified for *g*.

The general hierarchical model specified above often leads to analytically intractable computations of posterior distributions $p(\beta \mid Y)$ when estimating β. Approximate methods can be used for parameter estimation and can include a semi-Bayes, empirical Bayes, or fully Bayesian approaches. In a semi-Bayes approach, the value for τ^2 is prespecified as opposed to estimating the value from the data. This may be advantageous if the estimate of τ^2 is itself highly unstable (Greenland, 1993, 1997; Greenland and Poole, 1994). However, because the most appropriate value for τ^2 is unknown it is standard to perform a sensitivity analysis to evaluate how dependent the posterior estimates are to the choice of τ^2. In contrast, an empirical Bayes approach uses the marginal distribution of τ^2 to obtain a point estimate that is then used to evaluate the joint posterior distribution for the log odds ratios, β (Efron and Morris, 1975; Greenland and Poole, 1994; Morris, 1983; Searle *et al.*, 1992). Semi-Bayes and empirical Bayes methods, such as two-stage weighted least squares, joint iterative weighted least squares and penalized quasi-likelihood (Breslow and Clayton, 1993; Greenland, 1997), suffer from the inability to reflect our uncertainty for τ^2 in the final posterior estimates for the log of the odds ratio. A fully Bayesian approach using MCMC methods avoids this and incorporates the uncertainty about τ^2 into the analysis by evaluating the full joint posterior distribution through simulation. However, these methods contain many potential difficulties and care must be taken when implementing and interpreting the final results (Gelman, 1995; Gilks *et al.*, 1996).

Within the hierarchical modeling framework, connections to model selection procedures can be easily made by assuming a mixture model for $p(\beta \mid \theta)$. George and McCulloch (1993) introduced a Stochastic Search Variable Selection

(SSVS) algorithm by introducing a latent variable, $W_j = 0$ or 1, indicating whether term j is included in the model with a mixture prior for the coefficients:

$$p(\beta_j|\tau,\psi) = \begin{cases} N(0,\tau^2), & \text{if } W_j = 0 \\ N\left(0,(\psi\tau)^2\right), & \text{if } W_j = 1 \end{cases}$$

Here, ψ is a variance inflation factor defining the separation between two normal distributions centered at zero. Since τ^2 functions as a smoothing parameter, a specification of $\tau^2 = 0$ defines a point mass at zero for those terms not included in the model (see "spike and slab" discussion in Section IV).

The benefits of a Bayesian approach to parameter specification need not be limited to the case of numerous factors. Consider a simple scenario in which the aim is to estimate a statistical or multiplicative interaction between two factors, a genetic factor G and a dichotomous environmental factor E. In the conventional single-level case-control analysis, the departure from a multiplicative interaction model can be estimated as ${\beta_{GE}}^{CC}$ from the following logistic model:

$$\text{logit}(p(Y=1|G,E)) = \beta_0^{CC} + \beta_G^{CC}G + \beta_E^{CC}E + \beta_{GE}^{CC}GE$$

Under the assumption of a rare disease and independence of the two factors in the source population, a case-only analysis may be used to estimate an equivalent interaction ${\beta_{GE}}^{CO}$, as well:

$$\text{logit}(p(E=1|Y=1,G)) = \beta_0^{CO} + \beta_{GE}^{CO}G$$

Leveraging these two approaches, Mukherjee and Chatterjee (2008) constructed a hierarchical prior for the case-control estimate as a function of the case-only estimate, $\beta_{GE}^{CC} \sim N(\beta_{GE}^{CO},\tau^2)$. The result is a shrinkage estimator similar to that outlined above but one that is a weighted average between the case-only and case-control estimates:

$$\tilde{\beta}_{GE} = B\beta_{GE}^{CO} + (1-B)\beta_{GE}^{CC}$$

The weight B is defined as:

$$B = \frac{\hat{V}_{CC}}{\hat{V}_{CC} + \theta_{GE}^2}.$$

The amount of shrinkage is controlled by ${\theta_{GE}}^2$, the maximum likelihood estimate of the log of the G–E odds in controls relative to the estimated variance $\left(\hat{V}_{CC}\right)$ of the case-control estimator ${\beta_{GE}}^{CC}$. Thus, the final estimate $\tilde{\beta}_{GE}$ shares the efficiency of the case-only estimate when the two factors are estimated to be independent within the controls and the robustness of the case-control estimate otherwise.

Further demonstrating the close connection between hierarchical mixture models and Bayesian model averaging, Li and Conti (2009) proposed a similar weighted average. However, in their Bayesian model averaging approach, the weight is defined as $B = p(M_{CO} \mid Y)$, the posterior probability of the case-only model relative to the case-control model. This is calculated as:

$$p(M_{CO}|Y) = \frac{p(Y|M_{CO})p(M_{CO})}{p(Y|M_{CO})p(M_{CO}) + p(Y|M_{CC})p(M_{CC})},$$

where $p(M_\gamma)$ is the prespecified or semi-Bayes prior probability for model γ (either the case-only or case-control models) and $p(Y \mid M_\gamma)$ is the integration of the likelihood of model M_γ, estimated through a Laplace transformation. For a comparable likelihood between the two models they frame both models in a log-linear framework. Similar to the approach of Mukherjee and Chatterjee (2008), since the likelihood of the case-control model is a function of the estimated G–E association in the controls, θ_{GE} plays an important role in determining the weight.

IV. MODEL UNCERTAINTY

Model selection is the process of combining data and prior information to select among a group of statistical models $M_\gamma \in M$. In building a model, decisions to include or exclude covariates as well as uncertainty in how to code the covariates in the design matrix X_γ for any given model M_γ are based both on the prior hypotheses and the data. With many potential covariates, these decisions become difficult. Some algorithms will select variables to be included in the model, but only return a single "best" model (e.g., stepwise regression). These methods fail to account for model uncertainty, that is, a number of models may fit the data equally well. In a Bayesian framework, model uncertainty can be addressed by basing inference on the posterior distribution of models.

One simple example of the model uncertainty framework is in variable selection, where each model is defined by a distinct subset of p covariates and is specified by the indicator vector γ which is comprised of a set of p 0s or 1s indicating the inclusion or exclusion of each of the covariates in model M_γ. Then in the logistic regression framework each model is defined as:

$$\text{logit}\big(\text{p}(Y = 1|X, M_\gamma)\big) = \beta_0 + X_\gamma \beta_\gamma$$

where X_γ is the design matrix that is made up of the p_γ covariates in model M_γ, and β_γ is a p_γ dimensional vector of model-specific regression coefficients. Here, the model space $M_\gamma \in M$ is made up of 2^p possible models.

In general, the main quantity of interest that needs to be calculated is the posterior probability for any model M_γ given by the equation:

$$p(M_\gamma|Y) = \frac{p(Y|M_\gamma)p(M_\gamma)}{\sum p(Y|M_\gamma)p(M_\gamma)}$$

where $p(Y|M_\gamma)$ is the marginal likelihood, $p(M_\gamma)$ is the prior probability for a particular model, and the denominator is a constant found by summing over the entire model space. When p is small, one can exhaustively visit all the possible models. For p covariates, there are $t=2^p-1$ possible terms (including main effect and interaction terms), and 2^t possible models. As p increases the model space M quickly outgrows what is computationally feasible, and an approximation of the posterior distribution of models by MCMC methods is required. When carefully designed, these approaches can efficiently search through the model space.

Bayesian approaches introduce an additional layer of uncertainty to the model by specifying priors on the model space itself $p(M_\gamma)$. By Ockham's razor, the simplest models that explain our observations are preferred. This guidance can be formalized as a prior (Jefferys and Berger, 1991). Many approaches adopt a "spike and slab" prior distribution, where most regression coefficients are zero (the spike) and a few coefficients have some effect (slab) (Mitchell and Beauchamp, 1988). Others incorporate some form of lasso (lease absolute shrinkage and selection operator), which either shrinks coefficients or sets them to zero (Tibshirani, 1996). Some approaches directly penalize model complexity. Chen and Chen (2008) introduced a penalty term to the Bayesian information criterion (BIC) based on the size of the model space with the same number of variables as the current model. Wilson *et al.* (2010) introduced a Beta-Binomial prior on model size that holds the prior odds of any association constant as p increases, thus limiting false discoveries.

Biological knowledge can also be incorporated into a prior on the model space. This knowledge can be used as a prior on the probability that a coefficient is involved ($p(\gamma_j=1)$) and the effect given that it is involved. As discussed earlier, SSVS achieves this using a mixture prior for variable j, such that:

$$\beta_j = \gamma_j N(\mu_j, \tau_j) + (1-\gamma_j)\delta(0)$$

where $\delta(0)$ is the spike, and $N(\mu_j, \tau_j)$ is the prior on the mean and variance of β_j given that it is not zero (Conti *et al.*, 2003, 2009). Biological knowledge can also form the basis for model priors. For example, a hypothesized biological pathway can be used as a reference "prior topology," and structures closer to this reference have greater prior probability (Baurley *et al.*, in press).

MCMC methods are extremely flexible. Here, we describe the general design of the MH algorithm. A random walk MH algorithm explores models in the neighborhood of the current model (Robert and Casella, 2004). A random change to the current model M_{t-1} is performed to create the proposed model M'.

The changes allowed are often specific to the model structure. For instance, a change to a logistic regression model could be the addition or removal of a regression term. A new model is accepted as M_t with probability,

$$\alpha = \min\left(1, \frac{p(Y|M')p(M')q(M_{t-1}|M')}{p(Y|M_{t-1})p(M_{t-1})q(M'|M_{t-1})}\right)$$

where $p(Y \mid M')$ is the marginal likelihood of the model, $p(M')$ is the model prior, and $q(\ \mid\ )$ is the proposal density. There are many choices for the model prior and proposal density, and these influence the performance and behavior of the algorithm. After convergence, models are sampled from the posterior distribution of models.

For model selection applications where p is large, MCMC algorithms may have difficulties traversing a multimodal posterior distribution and poor efficiency arising from the scale of the model space. Recently, new samplers have been introduced that overcome some of these issues by exploiting multiple chains running in parallel. Evolutionary stochastic search (ESS) utilizes a "population" of MCMC chains operating at different temperatures, with hotter chains moving about the model space quicker than cooler chains. The chains are updated with local and global moves. Local moves explore models in the neighborhood of the current model (i.e., random walk) whereas global moves allow a chain to move to a new area of the model space by swapping states with another chain. This algorithm has been incorporated into the MISA (Multilevel Inference of SNP Associations) framework that computes posterior probabilities and Bayes factors (BFs) at the SNP, gene, and global level (Wilson *et al.*, 2010).

In a different strategy, parallel chains are utilized to tune the MCMC proposal density to better approximate the posterior density. This improves efficiency because less time is spent proposing models with little evidence. The methodology (known as PEAK) organizes the model space into subspaces linked through a graph (Baurley *et al.*, 2009). This graph can be informative, meaning it is derived from an ontology or domain expert or simply symmetric (a divide and conquer approach to the model space). The chains running on smaller model spaces tune the proposal densities for chains operating on larger spaces. The method has been applied to a childhood asthma case-control dataset and discovered several oxidative stress genes and gene–gene interactions for further investigation.

We can also allow for model uncertainty when incorporating interactions of covariates in the structure of the interactions and the covariates being included in the interaction. For instance, in logic regression, the model is of the form,

$$\text{logit}(p(Y = 1|X)) = \beta_0 + \sum \beta_j f_j(X_j)$$

where $f_j(X_j)$ is a Boolean combination of the risk factors (called a logic tree) and β_j is the net effect of that Boolean combination (Ruczinski *et al.*, 2003). The model allows for L logic trees, each containing observed variables and logical operators (AND, OR, and NOT) that can represent different types of interactions. Baurley *et al.* (2010) extended this framework to continuous variables where the function $f_j(X_j)$ became the net effect of a pathway structure (called a topology) such that,

$$f_j(X_j) = \theta_{n,1}\text{par}_1(X_n) + \theta_{n,2}\text{par}_2(X_n) + (1 - \theta_{n,1} - \theta_{n,2})\text{par}_1(X_n)\text{par}_2(X_n)$$

where $\text{par}_1(X_n)$ and $\text{par}_2(X_n)$ return the values of the parents of X_n in the topology (either an observed variable Z or a latent variable X) and $\theta_{n,1}$ and $\theta_{n,2}$ are parameters that can represent a range of interaction types.

V. DETERMINING NOTEWORTHINESS

Given the above defined models, we are interested in addressing two questions: (1) globally, is there an association between any of our covariates of interest and the outcome? and (2) if there is a global association, which individual covariates or structures of covariates are most likely driving the association?. Both of the questions can be answered via multilevel posterior probabilities (or conditional probabilities) and Bayes factors.

A. Global posterior quantities

We are first interested in addressing the following global hypotheses:

H_A: *At least one covariate is associated with the outcome of interest.*

H_0: *There is no association between the covariates of interest and the outcome.*

The extent to which the data supports each of the hypotheses is calculated via posterior probabilities $p(H_A \mid Y)$:

$$p(H_A|Y) = \frac{p(Y|H_A)p(H_A)}{p(Y|H_A)p(H_A) + p(Y|H_0)p(H_0)},$$

and $p(H_0 \mid Y) = 1 - p(H_A \mid Y)$. These quantities are a function of both the marginal likelihood of the hypotheses ($p(Y \mid H_A)$ and $p(Y \mid H_0)$) and the prior distributions placed on the hypotheses ($p(H_A)$ and $p(H_0)$). In particular, in the Bayesian variable selection framework the posterior probability of the alternative hypothesis that at least one covariate is associated takes on the form:

$$p(H_A|Y) = \sum_{M_\gamma \neq M_0} p(M_\gamma|Y),$$

where we simply sum up the posterior probability for all non-null models ($M_\gamma \neq M_0$). Also, the posterior probability of the null hypothesis takes on the form:

$$p(H_0|Y) = p(M_0|Y).$$

Decisions based on which hypothesis is more likely can then be made based on the posterior probabilities. These posteriors have an intuitive interpretation of the probability of the hypothesis conditional upon seeing the data.

Given that there is posterior evidence of the global hypothesis that at least one of the covariates of interest is associated with the outcome, we are further interested in answering the question of which covariate is most likely driving the association. This question can be answered based on marginal posterior probabilities. In particular, in the Bayesian variable selection framework, the posterior probability of any covariate X_i being associated can be calculated as:

$$p(\gamma_i = 1) = \sum_{M_\gamma \in M:\gamma_i=1} p(M_\gamma|Y),$$

which is simply the sum of the posterior probability for every model that includes the covariate X_i.

One should note that in case of calculating inclusion probabilities for highly correlated covariates (i.e., SNPs in LD) there is an expected dilution in the corresponding posterior probabilities due to the covariates providing competing evidence for an association and therefore the posterior probability of an association will be diluted or distributed across several correlated covariates. We therefore extend this notion of marginal covariate inclusion probabilities to group inclusion probabilities (where we assume less correlation will exist across the groups) and achieve multilevel posterior probabilities (Wilson *et al.*, 2010) by considering the posterior probability that at least one covariate within a given group is associated. One example would be in genetic association studies to group SNPs according to their corresponding gene and calculate gene inclusion probabilities that are simply the sum of the posterior probability of all models that include at least one of the SNPs within the given gene.

In the Bayesian framework, we can also calculate the ratio of the weight of evidence for any two hypotheses (H_A *vs.* H_0) based on BFs:

$$BF[H_A : H_0] = \frac{p(H_A|Y)}{p(H_0|Y)} \bigg/ \frac{p(H_A)}{p(H_0)}.$$

A BF (Kass and Raftery, 1995) compares the posterior odds of any two hypotheses to the prior odds and measures the change of evidence provided by data for one hypothesis to the other. Goodman (1999) and Stephens and Balding (2009) provide a discussion of the usefulness of BFs in the medical context and Wakefield (2007) and Whittemore (2007) illustrate their use in controlling false discoveries in genetic epidemiology studies. Jeffreys (1961) presents a descriptive

Table 3.1. Jeffreys Grades of Evidence (Jeffreys, 1961)

Grade	BF(H_A:H_0)	Evidence against H_0
1	1–3.2	Indeterminate
2	3.2–10	Positive
3	10–31.6	Strong
4	31.6–100	Very strong
5	>100	Decisive

classification of BFs into "grades of evidence" (reproduced in Table 3.1) to assist in their interpretation, which is also reproduced in the work of Kass and Raftery (1995). Thus, decisions about which hypothesis are more likely can be made based on these grades of evidence.

Jeffreys (1961) was well aware of the issues that arise with testing several simple alternative hypotheses against a null hypothesis, noting that if one were to test several hypotheses separately, then by chance one might find one of the BFs to be less than one even if all null hypotheses were true. He suggested that, in this context, the BFs needed to be "corrected for selection of hypotheses." However, it was not clear what Jeffreys meant explicitly by this correction. Experience with genetic studies shown that detectable SNP associations are relatively infrequent. For this reason, Stephens and Balding (2009) suggest that marginal BFs calculated assuming equal prior odds should alternatively be interpreted in light of prior odds more appropriate to the study at hand (leading to a much greater significance threshold than Jeffreys suggests). Another approach to the problem of exploring multiple hypotheses is to embed each of the potential submodels (corresponding to a subset of SNPs) into a single hierarchical model. Unlike the marginal (one-at-a-time) BFs indicated by Stephens and Balding (2009) that are independent of the prior odds on the hypotheses, SNP BFs computed in the Bayesian variable selection framework are based on comparing composite hypotheses and hence do depend on the prior distribution over models. Thus, it is important to select priors on the model space that have an implicit multiplicity correction. One example of this is the Beta-Binomial prior with hyperparameters $a=1$ and $b=p$ (the number of SNPs in the study) suggested by Wilson *et al.* (2010). By diluting the prior marginal inclusion probability, this maintains constant global prior odds of an association even as the number of SNPs in the analysis increases.

VI. CONCLUSIONS

Bayesian approaches to complex analysis are becoming more and more popular. In this chapter, we have attempted to demonstrate that a Bayesian perspective provides a flexible framework for complex genetic analyses by breaking the problem into several components: (1) specification of a model structure via the covariates or combination of covariates; (2) estimation and prior specification of the corresponding parameters; and finally, (3) incorporation of the uncertainty of the specified model structure. The themes presented here focus on a generalization of Bayesian hierarchical models and their flexibility to allow the analyst to easily incorporate complex structures, multiple parameters, deal with nuisance parameters, and have common sense interpretations of the parameters of interest. Unfortunately, this added flexibility over more commonly used frequentist methods comes with the added complexity of computing conditional probabilities (high-dimensional integrals or summations) and eliciting subjection or, when there is a lack of prior knowledge, developing objective prior distributions. However, in this chapter, we discuss advances that have been made in both estimating high-dimensional integrals and in the elicitation of prior distributions.

The most notable advantage of a hierarchical perspective to data analysis is that each stage becomes relatively easy to construct, understand, and interpret. Upon aggregating these stages, the overall model can be quite complex. However, even in the face of these complexities, inference is feasible by leveraging the specified hierarchy. The potential advantages of hierarchical models are often contingent upon the ability of the model, both the individual stages and the overall probability model, to provide an accurate representation of the true data generating mechanism. The potential fear that the prior will overwhelm the data has potentially lead many to shun Bayesian approaches. While all analytic models make some level of assumptions, it is important to understand that the specification of priors is not solely subjective. A prior on parameters simply specifies or structures an exchangeable class of parameters. It does not prespecify the degree to which those parameters may differ. Further uncertainty is incorporated by also searching over alternative models. Sensitivity analysis should be performed to gauge the dependence of final inference upon the prior structure. However, the goal of such sensitivity analysis should be to better understand the balance between the data and the prior. It should not be done to ensure that final inference is reflective of the data only and not the prior. After all, our goal is to leverage the hierarchy and the external information to gain potential advantages with inference. In complex systems analysis of biological processes, we believe that a Bayesian perspective via hierarchical models is appropriate since these processes often follow a conceptual hierarchy, that is, SNPs within genes, genes within biochemical pathways, pathways within physiological processes, physiological processes within social networks. Furthermore, recent advances in technology now make it feasible to generate massive amounts of data aimed at measuring the elements in this hierarchy (i.e., genomics, proteomics, metabolomics). To integrate such data will require complex models that are reflective of the overall process, but simplistic and intuitive at each stage of construction.

References

Baurley, J. W., Conti, D. V., Gauderman, W. J., and Thomas, D. C. (2010). Discovery of complex pathways from observational data. *Stat. Med.* **29**(19), 1998–2011. [PMID: 2922970].

Best, N. G., Tan, K. K., *et al.* (1995). Estimation of population pharmacokinetics using the Gibbs sampler. *J. Pharmacokinet. Biopharm.* **23**(4), 407–435. PM ID: 8882748.

Bois, F. Y. (2001). Applications of population approaches in toxicology. *Toxicol. Lett.* **120**(1–3), 385–394. PM ID: 11323198.

Breslow, N. E., and Clayton, D. G. (1993). Approximate inference in generalized linear mixed models. *J. Am. Stat. Assoc.* **88,** 9–25. PM ID.

Chen, J. H., and Chen, Z. H. (2008). Extended Bayesian information criteria for model selection with large model spaces. Biometrika **95**(3), 759–771. [PMID not indexed in PubMed]

Chen, L. S., Hutter, C. M., Potter, J. D., Liu, Y., Prentice, R. L., Peters, U., and Hsu, L. (2010). Insights into colon cancer etiology via a regularized approach to gene set analysis of GWAS data. *Am. J. Hum. Genet.* **86**(6), 860–71. [PMID 20560206]

Clewell, H. J., Andersen, M. E., *et al.* (2002). A consistent approach for the application of pharmacokinetic modeling in cancer and noncancer risk assessment. *Environ. Health Perspect.* **110,** 85–93. PM ID.

Conti, D., and Gauderman, W. (2004). SNPs, haplotypes, and model selection in a candidate gene region: The SIMPle analysis of multilocus data. *Genet. Epidemiol.* **27,** 429–441. PM ID.

Conti, D. V., and Witte, J. S. (2003). Hierarchical modeling of linkage disequilibrium: Genetic structure and spatial relations. *Am. J. Hum. Genet.* **72**(2), 351–363. PM ID: 12525994.

Conti, D. V., Cortessis, V., *et al.* (2003). Bayesian modeling of complex metabolic pathways. *Hum. Hered.* **56**(1–3), 83–93. PM ID: 14614242.

Conti, D. V., Lewinger, J. P., *et al.* (2009). Using ontologies in hierarchical modeling of genes and exposures in biologic pathways. *In* "Phenotypes and Endophenotypes: Foundations for Genetic Studies of Nicotine Use and Dependence, Vol. 20" (G. E. Swan, ed.), pp. 539–584. NCI Tobacco Control, Bethesda, MD.

Cordell, H. J. (2009). Detecting gene–gene interactions that underlie human diseases. *Nat. Genet.* **10,** 392–404. PM ID.

Cortessis, V., and Thomas, D. C. (2003). Toxicokinetic genetics: An approach to gene–environment and gene–gene interactions in complex metabolic pathways. *In* "Mechanistic Considerations in the Molecular Epidemiology of Cancer" (P. Bird, P. Boffetta, P. Buffler, and J. Rice, eds.). IARC Scientific Publications, Lyon, France.

Davidian, M., and Gallant, A. R. (1992). Smooth nonparametric maximum likelihood estimation for population pharmacokinetics, with application to quinidine. *J. Pharmacokinet. Biopharm.* **20**(5), 529–556. PM ID:.

Efron, B., and Morris, C. (1975). Data analysis using Stein's estimator and its generalizations. *J. Am. Stat. Assoc.* **70,** 311–319. PM ID:.

Gelman, A. (1995). Bayesian Data Analysis Chapman & Hall, London/New York.

Gelman, A., Bois, F., *et al.* (1996). Physiological pharmacokinetic analysis using population modeling and informative prior distributions. *J. Am. Stat. Assoc.* **91,** 1400–1412. PM ID:.

George, E. I., and McCulloch, R. E. (1993). Variable selection via Gibbs sampling. *JASA* **88,** 881–889. PM ID:.

Gilks, W., and Richardson, S. *et al.* (eds.) (1996). *In* Markov Chain Monte Carlo in Practice. Chapman & Hall, London.

Goodman, S. N. (1999). Toward evidence-based medical statistics. 2: The Bayes factor. *Am. Soc. Intern. Med.* **130,** 1005–1013. PM ID:.

Greenland, S. (1993). Methods for epidemiologic analyses of multiple exposures: A review and comparative study of maximum-likelihood, preliminary testing, and empirical-Bayes regression. *Stat. Med.* **12,** 717–736. PM ID.

Greenland, S. (1997). Second-stage least squares versus penalized quasi-likelihood for fitting hierarchical models in epidemiologic analyses. *Stat. Med.* **16,** 515–526. PM ID.

Greenland, S. (2000a). Principles of multilevel modelling. *Int. J. Epidemiol.* **29**(1), 158–167PM ID.

Greenland, S. (2000b). When should epidemiologic regressions use random coefficients? *Biometrics* **56**(3), 915–921. PM ID.

Greenland, S., and Poole, C. (1994). Empirical-Bayes and semi-Bayes approaches to occupational and environmental hazard surveillance. *Arch. Environ. Health* **49**(1), 9–16. PM ID.

Greenland, S., and Robins, J. M. (1991). Empirical-Bayes adjustments for multiple comparisons are sometimes useful. *Epidemiology* **2,** 244–251. PM ID.

Hung, R. J., Brennan, P., *et al.* (2004). Using hierarchical modeling in genetic association studies with multiple markers: Application to a case-control study of bladder cancer. *Cancer Epidemiol. Biomarkers Prev.* **13**(6), 1013–1021. PM ID 15184258.

Jefferys, W. H., and Berger, J. O. (1991). Sharpening Ockham's Razor on a Bayesian Strop. Technical Report #91-44C. Department of Statistics, Purdue University.

Jeffreys, H. (1961). Theory of Probability. Oxford University Press, Oxford.

Kass, R. E., and Raftery, A. E. (1995). Bayes factors. *J. Am. Stat. Assoc.* **90,** 773–795. PM ID.

Li, D., and Conti, D. V. (2009). Detecting gene–environment interactions using a combined case-only and case-control approach. *Am. J. Epidemiol.* **169**(4), 497–504. PM ID: 19074774.

Li, W., and Reich, J. (2000). A complete enumeration and classification of two-locus disease models. *Hum. Hered.* **50,** 334. PM ID.

Lunn, D., Best, N., *et al.* (2009). Combining MCMC with 'sequential' PKPD modelling. *J. Pharmacokinet. Pharmacodyn.* **36**(1), 19–38. PM ID: 19132515.

Mitchell, T. J., and Beauchamp, J. J. (1988). Bayesian variable selection in linear regression. *J. Am. Stat. Assoc.* **83**(404), 1023–1032. PM ID.

Moore, J. H. (2003). The ubiquitous nature of epistasis in determining susceptibility to common human diseases. *Hum. Hered.* **56**(1–3), 73–82. PM ID: 14614241.

Moore, J. H., and Williams, S. M. (2009). Epistasis and its implications for personal genetics. *Am. J. Hum. Genet.* **85**(3), 309–320. PM ID: 19733727.

Moore, J. H., Barney, N., *et al.* (2007). Symbolic modeling of epistasis. *Hum. Hered.* **63**(2), 120–133. PM ID: 17283441.

Morris, C. (1983). Parametric empirical Bayes inference: Theory and applications (with discussion). *J. Am. Stat. Assoc.* **78,** 47–65. PM ID.

Mukherjee, B., and Chatterjee, N. (2008). Exploiting gene-environment independence for analysis of case-control studies: An empirical Bayes-type shrinkage estimator to trade-off between bias and efficiency. *Biometrics* **64**(3), 685–694. PM ID: 18162111.

Park, T., and Casella, G. (2008). The Bayesian Lasso. *J. Am. Stat. Assoc.* **103**(482), 681–686PM ID.

Racine-Poon, A., and Wakefield, J. (1998). Statistical methods for population pharmacokinetic modelling. *Stat. Methods Med. Res.* **7**(1), 63–84. PM ID: 9533262.

Rebbeck, T. R., Martinez, M. E., *et al.* (2004). Genetic variation and cancer: Improving the environment for publication of association studies. *Cancer Epidemiol. Biomarkers Prev.* **13**(12), 1985–1986. PM ID.

Robert, C. P., and Casella, G. (2004). Monte Carlo Statistical Methods. Springer, New York.

Robins, J. M., and Greenland, S. (1986). The role of model selection in causal inference from nonexperimental data. *Am. J. Epidemiol.* **123**(3), 392–402. PM ID.

Rothman, K. J., and Greenland, S. (1998). Modern Epidemiology. Lippencott-Raven, Philadelphia.

Ruczinski, I., Kooperberg, C., *et al.* (2003). Logic regression. *J. Comput. Graph. Stat.* **12,** 475–511. PM ID.

Searle, S. R., Casella, G., *et al.* (1992). Variance Components. Wiley, New York.

Sebastiani, P., Ramoni, M. F., *et al.* (2005). Genetic dissection and prognostic modeling of overt stroke in sickle cell anemia. *Nat. Genet.* **37**(4), 435–440. PM ID: 15778708.

Sorenson, D., and Gianola, D. (2002). Likelihood, Bayesian and MCMC Methods in Quantitative Genetics. Springer, New York.

Speigelhalter, D. J., Thomas, A., *et al.* (2003). WinBUGS Version 1.4 User Manual Medical Research Council Biostatistics Unit, Cambridge.

Stephens, M., and Balding, D. J. (2009). Bayesian statistical methods for genetic association studies. *Nat. Genet.* **10,** 681–690. PM ID.

Stram, D. O., Pearce, C. L., *et al.* (2003). Modeling and E-M estimation of haplotype-specific relative risks from genotype data for a case-control study of unrelated individuals. *Hum. Hered.* **55**(4), 179–190. PM ID.

Tang, W., Wu, X., *et al.* (2009). Epistatic module detection for case-control studies: A Bayesian model with a Gibbs sampling strategy. *PLoS Genet.* **5**(5), 1–17. PM ID.

Thomas, D. (2010b). Methods for investigating gene–environment interactions in candidate pathway and genome-wide association studies. *Annu. Rev. Public Health* **31**, 21–36. [PMID PMC2847610].

Thomas, D. (2010a). Gene-environment-wide association studies: emerging approaches. *Nat. Rev. Genet.* **11**(4), 259–272. [PMID 2891422].

Thomas, D. C., Siemiatycki, J., *et al.* (1985). The problem of multiple inference in studies designed to generate hypotheses. *Am. J. Epidemiol.* **122,** 1080–1095PM ID.

Thomas, D., Langholz, B., Clayton, D., Pitkaniemi, J., Tuomilehto-Wolf, E., and Tuomilehto, J. (1992). Empirical Bayes methods for testing associations with large numbers of candidate genes in the presence of environmental risk factors, with applications to HLA associations in IDDM. *Ann. Med.* **24,** 387–392PM ID.

Thomas, D. C., Witte, J. S., *et al.* (2007a). Dissecting effects of complex mixtures: Who's afraid of informative priors? *Epidemiology* **18**(2), 186–190. PM ID: 17301703.

Thomas, P. D., Mi, H., *et al.* (2007b). Ontology annotation: Mapping genomic regions to biological function. *Curr. Opin. Chem. Biol.* **11**(1), 4–11. PM ID: 17208035.

Thomas, D. C., Conti, D. V., *et al.* (2010). Use of pathway information in molecular epidemiology. *Hum. Genomics* **4**(1), 21–42PM ID.

Tibshirani, R. (1996). Regression shrinkage and selection via the Lasso. *J. R. Stat. Soc.* **58**(1), 267–288PM ID.

Wakefield, J. (1996). The Bayesian analysis of population pharmacokinetic models. *JASA* **91,** 62–75PM ID.

Wakefield, J. (2007). A Bayesian measure of the probability of false discovery in genetic epidemiology studies. *Am. J. Hum. Genet.* **81,** 208–227. PM ID.

Wakefield, J. C., Best, N. G., *et al.* (2000). Bayesian approaches to disease mapping. *In* "Spatial Epidemiology: Methods and Applications" (P. Elliot, J. Wakefield, N. Best, and D. Briggs, eds.), pp. 104–127. Oxford University Press, New York.

Wang, K., Li, M., *et al.* (2007). Pathway-based approaches for analysis of genomewide association studies. *Am. J. Hum. Genet.* **81**(6), PM ID: 17966091.

Whittemore, A. S. (2007). A Bayesian false discovery rate for multiple testing. *J. Appl. Stat.* **34**(1), 1–9PM ID.

Wilson, M. A., Iversen, E. S., *et al.* (2010). Bayesian model search and multilevel inference for SNP association studies. *Ann. Appl. Stat.* PM ID.

Zellner, A. (1986). On assessing prior distributions and Bayesian regression analysis with g-prior distributions. *In* "Bayesian Inference and Decision Techniques: Essays in Honor of Bruno de Finetti" (P. K. BGoel and A. Zellner, eds.), pp. 233–243. Elsevier, North-Holland.

Zhang, Y., and Liu, J. S. (2007). Bayesian inference of epistatic interactions in case-control studies. *Nat. Genet.* **39**(9), 1167–1172. PM ID.

Zhong, H., Yang, X., *et al.* (2010). Integrating pathway analysis and genetics of gene expression for genome-wide association studies. *Am. J. Hum. Genet.* **86**(4), 581–591. PM ID.

4

Multigenic Modeling of Complex Disease by Random Forests

Yan V. Sun
Department of Epidemiology, School of Public Health, University of Michigan, Ann Arbor, Michigan, USA

ABSTRACT

The genetics and heredity of complex human traits have been studied for over a century. Many genes have been implicated in these complex traits. Genome-wide association studies (GWAS) were designed to investigate the association between common genetic variation and complex human traits using high-throughput platforms that measured hundreds of thousands of common single-nucleotide polymorphisms (SNPs). GWAS have successfully identified many novel genetic loci associated with complex traits using a univariate regression-based approach. Even for traits with a large number of identified variants, only a small fraction of the

Advances in Genetics, Vol. 72 0065-2660/10 $35.00

DOI: 10.1016/S0065-2660(10)72004-6

interindividual variation in risk phenotypes has been explained. In biological systems, protein, DNA, RNA, and metabolites frequently interact to each other to perform their biological functions, and to respond to environmental factors. The complex interactions among genes and between the genes and environment may partially explain the "missing heritability." The traditional regression-based methods are limited to address the complex interactions among the hundreds of thousands of SNPs and their environmental context by both the modeling and computational challenge. Random Forests (RF), one of the powerful machine learning methods, is regarded as a useful alternative to capture the complex interaction effects among the GWAS data, and potentially address the genetic heterogeneity underlying these complex traits using a computationally efficient framework. In this chapter, the features of prediction and variable selection, and their applications in genetic association studies are reviewed and discussed. Additional improvements of the original RF method are warranted to make the applications in GWAS to be more successful.

DEFINITION

SNP — single-nucleotide polymorphism

GWAS — a genome-wide association study is an examination of the association between genetic variation across a whole genome and a given trait.

Complex Disease — a complex disease is caused by a combination of genetic, environmental, and lifestyle factors, most of which have not yet been identified.

Genetic Architecture — it refers to the complete genetic basis, which includes various types of genetic effects, underlying a given trait.

Ensemble Learning — ensemble learning is the process by which multiple models (learners) are strategically generated and combined to jointly solve a prediction problem.

Epistasis — gene–gene interactions. It refers to the effect of one gene is modified by one or several other genes.

Linkage Disequilibrium (LD) — LD is the nonrandom association of alleles at two or more loci.

I. INTRODUCTION

Many human traits and diseases are highly heritable, which indicates a strong genetic component associated with these traits and diseases. The genetic studies in human populations have estimated the heritability and identified genetic loci linked to these human traits and diseases. Recent advancement of biotechnology, particularly the microarray technology, allows hundreds of thousands of measurements of a biosample in a single assay. The low cost and robust measurement of this high-throughput technology enabled the genome-wide association studies (GWAS), scanning of genome-wide genetic variants associated with complex traits and diseases in large samples (Altshuler *et al.*, 2008; Frazer *et al.*, 2009). Although over 2000 genetic variants have been successfully identified associated with human diseases by GWAS using univariate approach (Hindorff *et al.*, 2009) (A Catalog of Published Genome-Wide Association Studies. Available at: www.genome.gov/gwastudies. Accessed April 15, 2010), only a small portion of the heritability of most complex diseases had been explained by these findings (Manolio *et al.*, 2009). The current approaches are limited to test the marginal effects of genetic variants, and little has been explored to uncover the complex genetic architecture underlying these diseases. Rather than functioning linearly and independently, genes and their protein products function as interactive complexes in biological pathways, networks, and systems. The genetic architecture involves these higher order genetic interactions and their relationship with the environmental factors. The concept of genetic architecture has been well developed and discussed (Flint and Mackay, 2009; Sing and Boerwinkle, 1987). Thorough application in human disease has not been conducted limited by the scale of the measurements before the GWAS era. Better understanding the genetic architecture of complex disease is feasible with the large amount of genotypic and phenotypic measurements on large samples from GWAS.

On the other hand, the large datasets from GWAS (millions of predictors and thousands of individuals) posed computational challenges in the data management and statistical analysis. Some of these challenges, especially the genetic analysis of the marginal effects and the pair-wise interaction effects, had been addressed (Aulchenko *et al.*, 2007; Purcell *et al.*, 2007). However, the modeling tools of higher order relationship among genetic variants beyond univariate association and pair-wise interaction have not been well represented. An exhaustive screening of higher order interaction effects (e.g., three way interactions) requires large amount of computational resources and stringent statistical threshold due to multiple testing. To search for additional genetics effects of complex diseases and to improve the prediction of these diseases, alternative methods were considered to complement the classic statistical genetic methods.

Distinguishing "disease" from "non-disease" can be regarded as a classification problem, or supervised learning, a major topic in the machine learning field. Many powerful methods have been developed such as classification and regression tree (CART), support vector machine (SVM), Random Forests (RF), Bagging and Boosting. Each of these approaches has some unique features that enable them to perform well in certain scenarios. On the other hand, all approaches are quite flexible and have been applied to an array of biomedical problems. Ensemble learning algorithms, such as Boosting, Bagging and RF, have emerged as important machine learning tools for classifying human diseases and diagnosis related phenotypes in high-dimensional biomedical data analyses. These ensemble learning methods share similar advantages of good predictability, insensitivity to outliers, limited effort of model tuning, stable performance and insensitivity to uninformative predictors (Hastie *et al.*, 2001). RF, as one of the recently developed machine learning methods, has been recognized as a useful tool for classification and prediction problems in biomedical research (Chen and Liu, 2005; Diaz-Uriarte and Alvarez de Andres, 2006).

Given that RF is a robust prediction method which can incorporate complicated relationships among predictors and can deal with high-dimensional data, RF has been applied to analyze transcriptomic (Huang *et al.*, 2005), proteomic (Izmirlian, 2004), epigenomic (Christensen *et al.*, 2009) as well as genetic data (Bureau *et al.*, 2005; Sun *et al.*, 2007, 2008). RF was regarded as a powerful tool to screen the genetic variants predicting the phenotypes, and to model the higher order interaction effects with the high-dimensional genetic data (Lunetta *et al.*, 2004). Recent developments of RF have addressed a series of issues specific to the genomic analysis of human diseases using genome-wide single-nucleotide polymorphism (SNP) data. These developments and the applications of RF in modeling genetic architecture of human disease are discussed in this chapter.

II. RANDOM FORESTS

GWAS were designed to investigate the association between common genetic variation and complex human traits using high-throughput platforms that measured hundreds of thousands of common SNPs in epidemiological samples. In the last 3 years, GWAS have successfully identified many novel genetic loci associated with complex human diseases and traits (Altshuler *et al.*, 2008; Frazer *et al.*, 2009). However, most of the identified genetic variants have small effect size (e.g., the odds ratios for the heterozygote genotypes of the associated variants are approximately 1.1). Even for traits with a large number of identified variants, only a small fraction ($< 10\%$) of the interindividual variation in risk phenotypes has been explained. Since the heritability of these traits is usually higher than 30–40%, it is speculated that the "missing heritability" can be partially explained by interaction effects among the genetic variants (Manolio *et al.*, 2009; Moore and Williams, 2009). The traditional regression-based methods are limited to address the complex interactions among the 100,000s of SNPs and their environmental context by both the modeling and computational challenge (Moore *et al.*, 2010). Several machine learning methods are regarded as useful alternatives to capture the complex interactions effects among the high-dimensional GWAS data and potentially address the genetic heterogeneity underlying these disease traits.

Supervised machine learning algorithms can search through a hypothetical space to find a set of variables modeled in a way to accurately predict the outcome. An accurate prediction model can be very difficult to find when the dimensionality of the search space is high and the variables have the complex relationship (e.g., nonlinear and interacting) with the outcome. Ensemble learning methods have been developed to identify the predictive model by combining the outputs of many weak learners to produce a powerful "committee" for prediction. An ensemble learning method is a supervised learning machine which can be trained and then predicts the outcome. The accurate prediction of ensemble learning methods is achieved by essentially eliminating the uncorrelated errors of individual learners by averaging (Hansen and Salomon, 1990). In order for the methods to be effective, the individual learner needs to represent certain degree of diversity among themselves. The diversity among the learners corresponds to the unrelatedness among the classifiers in the context of a classification problem. The diversity in the classifiers can be achieved by using different subsamples or different parameters of models for each classifier. This process enables individual classifiers to produce various boundaries in the hyperspace. These relatively unrelated classifiers give uncorrelated errors and a strategic ensemble reduces the overall error rates. These ensemble learning methods work the best for classification problems with complex, nonlinear boundary among the categories.

The procedure of ensemble learning is conceptually illustrated in Fig. 4.1. The procedure of ensemble learning generally consists of two phases, the model fitting phase and the prediction phase. Many individual learners (L_1 to L_m) are built and fitted to subsamples (S_1 to S_m) resampled from the original sample S. In the prediction phase, each learner (L_1 to L_m) gives its own

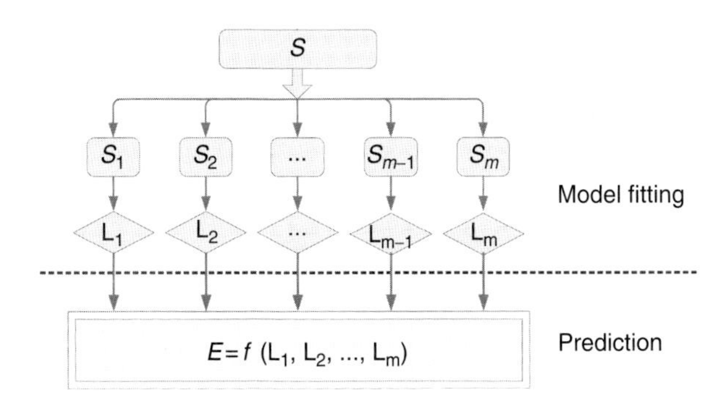

Figure 4.1. Conceptual model of ensemble learning methods.

prediction of a given input. The ensemble predictor E ("committee") combines all the individual predictions to make a final prediction of the input. Building and evaluating an ensemble learning model requires more computation than a single model prediction method, because many individual models need to be fitted and subsampling is usually required. However, ensemble learning methods have an advantage of being able to capture a more complex structure among the predictors where such structure does exist (e.g., biological pathways in the high-dimensional transcriptomic and genomic data). Superior performance of prediction is always more favorable than the extra computation cost in such studies where the single model algorithm does not predict well. Fast algorithms, such as classification and decision trees CART, are commonly used with ensemble learning methods to offer a computationally efficient ensemble learning package.

Ensemble learning methods strategically generate multiple models and combine them to solve a particular prediction problem. These methods are primarily used to improve the performance of prediction. They can also assist to select optimal set of variables for prediction or rank the variables by their contribution to prediction (i.e., so-called "variable selection"). Empirically, ensemble learning methods tend to yield better results when there is a significant diversity (i.e., heterogeneity) among the models (Kuncheva and Whitaker, 2003). This is a very interesting property of predicting a heterogeneous outcome. For complex traits, multiple molecular pathways, networks, and systems are involved in the process. The heterogeneous nature under these complex disease traits may cause the poor performance of single prediction model. The diversified models within the ensembles can very well match the heterogeneity of complex disease if the models are built appropriately. Thus, the ensemble learning methods are expected to provide superior performance of predicting these complex traits.

The RF is a powerful ensemble learning method developed by Leo Breiman (Breiman, 2001). The RF combines many classification and decision trees (CART) and is also distinct from other tree-based ensemble learning methods in how the trees are grown: (1) The method randomly selects, with replacement, n samples from the original training data (bootstrap with replacement); (2) At each splitting node, only a small group of input variables are randomly selected from all variables; (3) Each tree is grown to the largest extent possible until all the terminal nodes are maximally homogeneous.

The flowchart of RF algorithm is illustrated in Fig. 4.2. To grow a single CART tree, RF starts with creating a bootstrap sample (with replacement) from the original training set. Then a subset of predictors is randomly selected at each node to be split. A variable for a split is selected by its ability to improve the "purity" of the nodes. This node splitting process continues until all of purity measurements of the terminal nodes could not be improved. Comparing to CART, each tree in RF does not need to be pruned (Breiman *et al.*, 1984). This tree growing process repeats until a

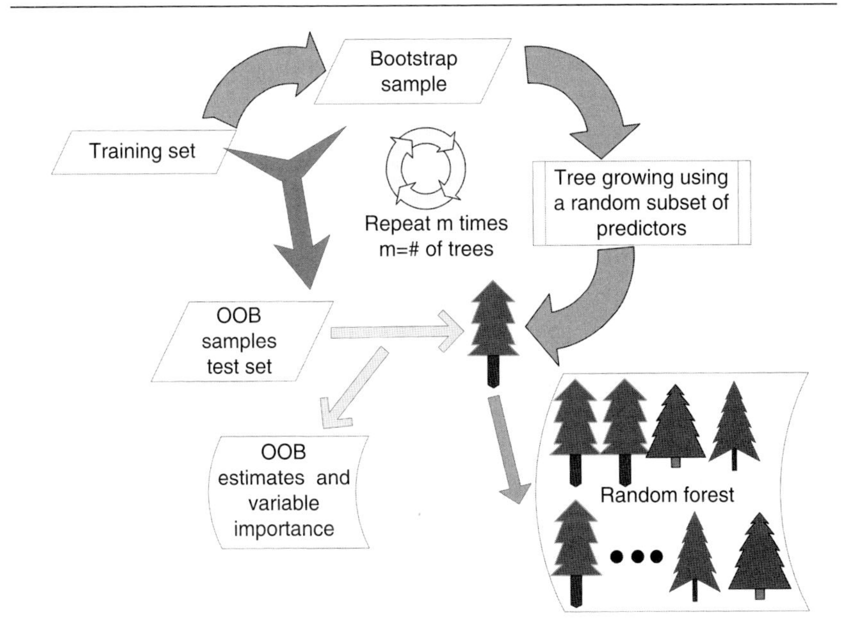

Figure 4.2. Flowchart of Random Forests.

preset number of trees is reached. All the trees in the forest are different in the sense of divergent variables at each split and divergent structure (topology) of the binary tree. The forest (i.e., ensemble of the trees) serves as the final machine for prediction.

For growing each tree in the forest, a bootstrap sample from the original data was selected. The remaining data are said to be "out-of-bag" (OOB) which is approximately one-third of the total samples. The OOB data can serve as a test set for the tree grown on the bootstrap sample. For a single sample in the training set, the probability of being OOB sample of a tree is about one-third (i.e., it is an OOB sample for about one-third of the trees). A tree can predict the outcome of the OOB samples and RF can predict the outcome by summarizing all individual trees. An OOB error rate is calculated by averaging all samples. In another word, the OOB error rate is the percentage of the time (individual tree prediction) that the RF predictor is correct for a classification problem. The OOB error rate can be used to optimize the parameters (e.g., mtry) of growing the forest by selecting a value in a range to minimizing OOB error rate. The OOB samples are also used by RF to estimate the permutation-based variable importance which is an important measurement of variable selection in RF (more details in Section IV.B). Since the OOB estimates are only based on about one-third of all trees, the OOB error rate tends to be overestimated (lower than it actually is). It is essential to run sufficient number of trees to make sure the test set error converges in order to have unbiased OOB estimate (Breiman, 2001).

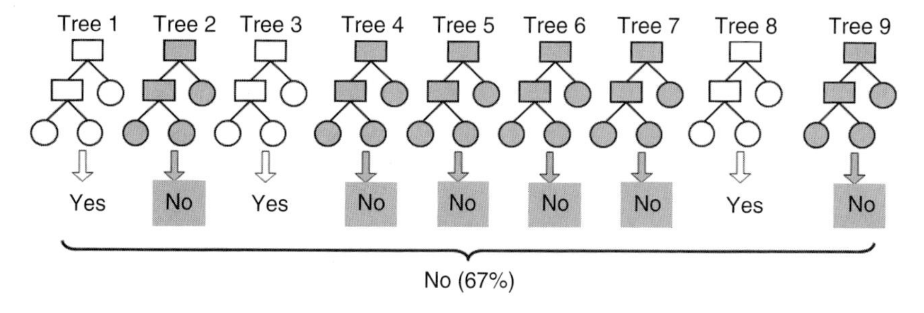

Figure 4.3. Committee voting in Random Forests.

To classify a new sample from an input, every single tree in the forest needs to be processed. Based on the classification (vote) from each tree, the forest chooses the classification having the most votes over all the trees in the forest. In a simplified case of nine trees in RF (Fig. 4.3), six of the nine trees classifies an input as "No" and three classifies as "Yes." As an ensemble, the final classification of this sample in RF is "No" which has a majority vote of 67%. Two types of randomness, bootstrap sampling and random selection of input variables, are implemented in the algorithm to make sure that the decision trees grown in the forest are dissimilar and uncorrelated from each other. Growing a forest of trees and using randomness in building each classification tree in the forest leads to better predictions in comparison to a single classification tree and helps to make the algorithm robust to noise in the dataset (Breiman, 2001). The RF enjoys several nice features: like boosting, it is robust with respect to input variable noise and overfitting, and it gives estimates of what variables are important in the classification.

The RF algorithm provides several measurements of estimating the relative importance of each variable on prediction. Therefore, it provides a way to select the most outcome-related variables for further verification studies. As one of the best prediction tools, RF method does not have the "overfitting" problem. These important features make the RF algorithm to be an outstanding candidate for analyzing high-dimensional data with many uninformative variables, such as high-throughput transcriptomic, genomic, and proteomic data (Chen and Liu, 2005; Diaz-Uriarte and Alvarez de Andres, 2006; Izmirlian, 2004; Qi *et al.*, 2005; Shi *et al.*, 2005; Wu *et al.*, 2003).

RF also inherited the capability of dealing with heterogeneous data from the ensemble learning algorithm. For complex disease traits such as hypertension, there are multiple genetic pathways and systems involved in long-term and short-term blood pressure regulation (Guyton *et al.*, 1972; Savoia and Schiffrin, 2006). RF is an appropriate candidate to capture the genetic heterogeneity underlying the trait because RF itself is an ensemble of many heterogeneous trees built from uncorrelated subsamples of the original dada. Two most widely applied utilities of RF in genetic research, prediction of the outcome and selection of genetic variants related to the outcome, as well as the interpretation of the RF results are discussed in the rest of this chapter.

III. PREDICTION

RF and other ensemble learning methods can be considered as a form of the Roman "divide and conquer" strategy. For a complex classification problem, defining a simple boundary is not able to separate all the different classes, especially in a high-dimensional space. Alternatively, we can focus on a subset of data (divide) to find a solution specific to this subset (conquer). By combining the individual classifiers and boundaries (ensemble such as RF), we are able to define a nonlinear, complex boundary to better separate the classes in the high-dimensional space. In RF, the "divide" phase is not limited to the samples. It extends to "dividing" the space by only selecting a subset of variables for growing each tree. Although each tree may only predict well for a subsample, it is expected to be able to partially predict the overall sample. Thus each individual tree is a weak learner. As previously described, these trees are also uncorrelated, and together they can average out the prediction errors.

Genetic variants represent the inter-individual differences on the DNA level. These variants in an individual's DNA are mostly inheritable and contributed to the heritability of many human traits. Genetic association studies have long been conducted to identify genetic variants linked to human traits and diseases. For complex disease, usually a number of biological processes can affect the trait at different organs, tissues, cellular compartments, and biochemical pathways. Therefore, it is very likely that individuals developed the same disease through the distinct genetic mechanisms. In the example of hypertension (high blood pressure), the genes in the rennin–angiotensin system, inflammation pathway, sodium channels, endothelin system, and sympathetic nervous system are all known to affect the regulation of blood pressure. Without knowing the subtypes which separate the patients, we lose the power to detect those genetic variants only affecting one disease subtype. The mainstream approach pools all the individuals in a single analysis and the traditional methods are not able to model or capture the genetic heterogeneity underlying the complex disease.

In addition to the feature of handling high-dimensional data and incorporating the interactions among predictions, the "divide and conquer" process implemented in RF is also well suited to model the genetic heterogeneity of complex traits. If the individual trees model some of the genetic pathways and systems and predict a subgroup of people with disease, RF as an ensemble can significantly improve the prediction of the disease. These appealing features of RF encouraged researches to apply RF in prediction problems in biomedical studies right after RF algorithm was developed. Two scenarios of predicting disease using genetic and genomic data are discussed in the following sections.

A. Predicting disease by genetic profile

In genetic association studies of complex disease, a simple application of RF is to classify the disease and nondisease samples based on the individual's genetic profile. Since the DNA sequences are essentially unchanged through the life-course, the disease risks coming from the variations in primary sequence of DNA can be robustly measured and be used as a tool for disease prevention, diagnosis, and treatment. If the genomic profile of an individual reliably predicts the disease risk, even a moderate increase of developing a certain disease may strongly motivate an individual to participate in the intervention program before a clinical symptom develops. This earlier intervention has stronger accumulative effects for chronic disease, which develops over a long term. The predictive genomic profile serves as a foundation for personalized or individualized medicine. Here, I only focused on searching for a predictive model of complex disease using genetic variants.

Before the GWAS era, genetics variants (e.g., SNPs and microsatellites) have been measured to study the genetic associations with complex diseases. Although the dimensions of these linkage or candidate gene data are much smaller than GWAS, the same principle of predicting disease applies. In an application of RF to classify the asthma cases by using 42 SNPs on gene ADAM33, a 44% misclassification rate was achieved by only using the SNPs (Bureau *et al.*, 2005). The misclassification rate is not only lower than the expected random misclassification rate of 50%, but also lower than the lowest misclassification rate of 47.8% among 100 permutation runs. Although clinically the predictive ability of RF model in this study is not practical, statistically the model can assist classifying the patient and control population.

The measurement of the area under curve (AUC) of Receiver Operating Characteristic (ROC) curve is a more robust estimate of the overall accuracy of classification, particularly when the samples in the two classes are not balanced. The AUC of ROC curve is not dependent on the arbitrary threshold set to define cases and controls in RF (Sun *et al.*, 2007, 2008). Instead of using a fixed threshold, the ROC curve defines all possible thresholds and calculates the sensitivity and specificity accordingly. It is also flexible to determine the threshold to achieve a balanced sensitivity and specificity. The AUC of ROC curve generated by RF provides an unbiased measurement of the performance of prediction across different models. In another genetic study of predicting coronary artery calcification (CAC) using 305 variables including 287 SNPs, a good predictive accuracy was achieved by RF model even without any tuning effort (Sun *et al.*, 2008). The prediction accuracy was significantly higher than random comparing to 1000 permutation tests.

B. Improvement of disease prediction by genetic profile

From epidemiological studies, many nongenetic risk factors have been identified. Age, gender, total serum cholesterol, HDL cholesterol, systolic blood pressure, and smoking status are associated with coronary heart disease (CHD). Framingham risk score of CHD, a simple linear combination of these risk factors, can predict an individual's risk of having CHD even over the next 10 years. Because its estimate is robust and all of these nongenetic risk factors are easy to measure, the Framingham risk score has been widely adopted. Although novel risk factors of CHD have been identified, the improvement of risk prediction of having CHD event has been marginal over decades. Based on the genetic risk factors identified in recent GWAS, Morrison *et al.* developed a genetic risk score (GRS) to predict CHD risks in both European Americans and African Americans (Morrison *et al.*, 2007). The increase in the AUC of ROC curve was statistically significant for African Americans and suggestive for European Americans, when the GSR was included with the traditional CHD risk score. This study demonstrated that the aggregating genetic risks from multiple SNPs improved prediction of CHD events. Given that the GRS was an additive combination of a small set of SNPs with marginal effects, the improvement of prediction through applying a better prediction model (e.g., RF) is promising. The contribution of GRS (18 genes) in improving disease classification was found to be minimal for type 2 diabetes (Talmud *et al.*, 2010). In a study of predicting cardiovascular disease, after adjustment for traditional cardiovascular risk factors, a GRS comprising 101 SNPs was not significantly associated with the incidence of cardiovascular disease in women (Paynter *et al.*, 2010). Sun *et al.* (2008) applied RF to classify CAC burden using both nongenetic and genetic risk factors. Although the overall model with both genetic and nongenetic factors predicted CAC measured by AUC of ROC curve, including the candidate gene SNPs did not improve the predictive ability significantly comparing to the nongenetic factor only model (unpublished data). Given that individual genetic factor usually has very small effect size and only a few hundred SNPs were modeled, this lack of improvement of predictive ability should not be generalized to the genome-wide SNP study. So far, there is no published result demonstrating if the genome-wide SNP data can significantly improve the predictive ability over the model of known nongenetic risk factors. By identifying and incorporating more genetic risk factors and their interactions using a RF model, the overall contributions of these genetic risk factors can be stronger and likely provide a better risk estimate than the traditional risk scores of complex diseases.

IV. VARIABLE SELECTION

The feature of variable selection in RF is highly integrated with the procedure of building the predictive model. The best predictors are ranked based on their involvement in predicting the outcome in the RF model, rather than a predefined statistical relationship. Therefore, a direct comparison between RF and the regression-based statistical tests can be misleading. Here, the utility of RF in identification of genetic predictors of complex diseases, and the modification of RF to address some unique characteristics of genetic data are discussed in the following three sections.

A. Data dimensionality and search space

RF is known to be able to handle "high-dimensional" data (Breiman, 2001). The term "high-dimensional" can be translated to an arbitrary number by various researchers. It is particularly ambiguous for the genetic epidemiologist who used to deal with hundreds of genetic markers and currently deal with hundreds of thousands of common genetic variants in GWAS. When we expand the scope of GWAS to rarer genetic variants, the number is approximately 10 times larger. Is RF really capable of handling such "high-dimensional" data? There are two levels of issues, algorithm level and application level, related to this question.

Classic classification approaches tend to fail when $p \gg n$, where p is the number of predictors and n is the number of samples. RF had the comparable predictive ability to other classification methods, such as SVM and k-nearest neighbors, using microarray data measuring more than 2000 predictors (e.g., genes) on less than 100 samples, while selected a very small set of genes (Diaz-Uriarte and Alvarez de Andres, 2006). When the number of predictors is very large and the proportion of the informative markers is small, the performance of RF tends to decline. This observation was reported in both SNP analysis of rheumatoid arthritis (Sun *et al.*, 2007) and gene expression microarray analysis of mice (Amaratunga *et al.*, 2008).

In the study of rheumatoid arthritis, a total of 5742 autosomal SNPs were available on 740 unrelated individuals. Using the AUC of the ROC curve to represent the predictability of each RFs model, the overall predictability declined when more than 500 top ranking SNPs were used to build the model. When all the 5742 SNPs were included, the AUC of ROC dropped to 0.53 which suggested very poor predictive ability to classify the cases and controls. The predictive ability also dropped slightly when too few SNPs were used (ranging from 50 to 500). When random sets of SNPs instead of the top ranked ones were selected, the predictive ability of RF did not change and remained low (AUC of ROC was around 0.45–0.6). This pattern of decreased predictive ability of RF due to a large number of uninformative predictors is illustrated in Fig. 4.4. The performance of prediction peaks when the most informative markers are selected. Even RF was observed to be robust to uninformative predictors when the dimensionality was high, there is clearly a limit of the "high dimensionality." RF starts losing its predictive ability beyond the threshold and eventually unable to predict the outcome (the "decline stage" in Fig. 4.4). These observations suggested that (1) including too many noninformative predictions decreases the ability of RF to build an accurate prediction model; (2) selecting the most

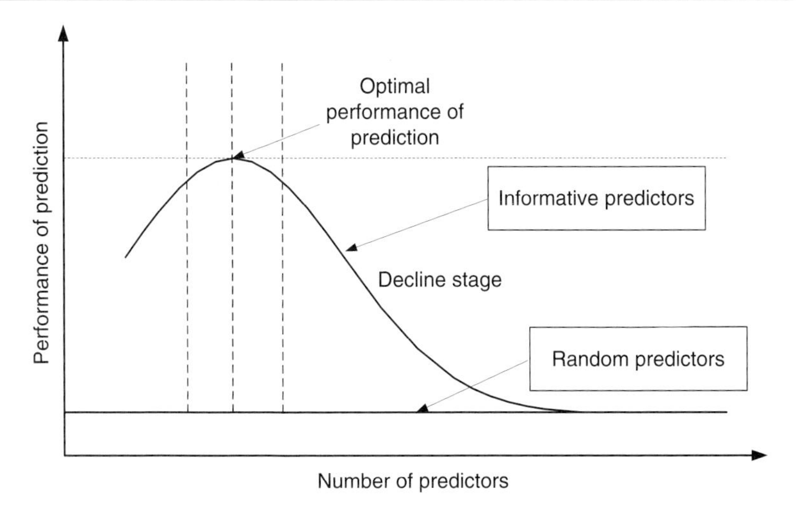

Figure 4.4. The impact of uninformative predictors on the prediction performance of Random Forests.

informative predictor among the large pool of predictors can improve the performance of RF. For an even higher dimensional GWAS data (usually over hundreds of thousands SNPs) with smaller proportion of informative predictors (<1%), selecting the most informative markers can be essential to build an accurate RF model.

This problem of a large number of noninformative predictors in RF is rooted in the way ensemble learning methods work. The ensemble learning methods achieve low error rate of prediction by averaging out the prediction errors of many individual learners. The overall prediction only improves when the individual learner also predicts, even slightly better than random. In the case of having a large number of noninformative predictors, most of the CARTs (the individual learner in RF) may be generated only by noninformative predictors and have no predictive ability of the outcome. Therefore, the RF, an ensemble of the CARTs, may have very poor power to predict or classify the outcome. Since RF has been demonstrated to be robust if the signal-to-noise ratio is not extremely small (Breiman, 2001), a prefiltering procedure may reduce the impact of the noninformative predictors by reducing the search space to those putatively informative predictors.

In a study of gene expression microarray of knockout mice, Amaratunga *et al.* chose the eligible subsets of predictors at each node of a tree by weighted sampling rather than random sampling, and demonstrated the superior performance of the modified RF approach, enriched RF (ERF) (Amaratunga *et al.*, 2008). The weight of ERF was based on q-values of false discovery rate (FDR) which controls for multiple testing of p-values. Comparing to the p-value-based weights (w_p), the q-value-based weights (w_q) are robust to those low p-values caused by chance from a large number of tests. Since the predictors would be assigned equal w_q in this case, the prediction model of ERF is less likely to be over-fit. The weight of the ith predictor, $w_{qi}=(1/q_i)-1$. In order to prevent a large proportion of predictors being assigned zero ($q_i=1$), or a small number of predictors having high weight (small $q_i>0$), the weights were restricted to a lower boundary a_{min}, and a higher boundary a_{max}. The adjusted weight of the ith predictor in ERF is w_i=median ($a_{min,}$ $w_{qi,}$ a_{max}), where $a_{min}=0.01$, $a_{max}=999$ (gives five-order dynamic range of variation). To address the issue of small sample size of most microarray studies, a balanced leave-one-out cross validation (LOOCV) procedure was tested against bagging. The LOOCV variant of ERF is computationally efficient since the weights are calculated N times (N is the sample size), instead of R times (the number of trees) for bagging. Using two gene expression microarray datasets in their knockout mice study and eight additional gene expression microarray datasets, the ERF outperformed RF, especially when the contract between the two categories of the outcome was subtle. The LOOCV variant of ERF was recommended due to its superior performance and better computational efficiency for datasets with small sample size.

The motivations of modifying RF in gene expression microarray study equally apply to prediction problems in GWAS, although the degree of impact of each factor may vary. For example, the dynamic range of a_{min} and a_{max} need to be bigger because a GWAS usually has much larger sample sizes (at least thousands) and a much larger number of predictors (hundreds of thousands) than gene expression microarray studies. The number of truly informative predictors and their overall predictive ability of the outcome are largely unknown. To identify additional genetic variants which can predict disease outcomes and traits, simulation studies on the GWAS scale are needed to estimate and evaluate the appropriate parameters for applying ERF in GWAS data.

The R package of RF, randomForest, has been widely used for classification problems of biomedical research. It offered the RF functions originally implemented in FORTRAN under the popular framework which can be easily integrated with a huge collection of R packages of data management, data analysis, and data visualization. Its computational limitation as an R function came to the surface when large GWAS datasets were applied. Schwarz *et al.* developed a stand-along version of RF in C++, which offers improved performance of managing and analyzing GWAS data (randomjungle.com) while retaining the basic functions of variable selection, missing value imputation, classifier creation, and prediction in RF. The continuous effort of implementing and improving RF in a fast and user-friendly form will greatly reduce the obstacles which prevent geneticists applying the advanced machine learning tools such as RF to uncover more genetics risk factors underlying complex diseases.

B. Potential biases of fitting genetic data

Two types of VIM are implemented in RF to estimate the importance of a variable in predicting the outcome, Gini importance and permutation importance. Gini VIM is a Gini index-based importance measurement inherited from the splitting rule of CART. In CART, Gini index measures the impurity of a node for a categorical outcome (Breiman *et al.*, 1984). If a node t containing samples from n classes, the Gini index is defined as

$$\text{gini}(t) = 1 - \sum_{j=1}^{n} p_j^2$$

where $p_j = n_j/n$, the relative proportion of category j in node t. If a dataset T is split into two subsets T_1 and T_2, with sizes of N_1 and N_2, respectively, the Gini index of the split is defined as $\text{gini}_{\text{split}}(T) = \frac{N_1}{N}\text{gini}(T_1) + \frac{N_2}{N}\text{gini}(T_2)$, a weighted average of the Gini index from the two subsets. The Gini index ranges 0–1−1/k, where 0 indicates a pure dataset with a single category, and 1−1/k represents an impure dataset with samples evenly distributed across k categories. In CART, a split s at node t is selected to maximize the improvement in Gini index of node t. The improvement in the Gini index of a binary split is defined by

$$\Delta I(s,t) = \text{gini}(t) - p_L \times \text{gini}(t_L) - p_R \times \text{gini}(t_R)$$

where t_L, t_R are the left and right child nodes of a node t, p_L and p_R are the proportion of samples in node t split into left and right child nodes (Fig. 4.5). The Gini index-based splitting rule looks at the largest category in the parent node and searching for a split to separate it from the other categories. An ideal series of splits would end up with k pure child nodes, one for each of the k categories in the target. For RF, this Gini index-based splitting rule is applied to grow each individual tree. Adding up the Gini improvement for each individual variable over all trees in the forest gives a fast variable importance. The Gini VIM of RF is often consistent with the permutation-based importance measure (www.stat.berkeley.edu/~breiman/RandomForests/).

The permutation VIM was considered a more advanced importance measure available in RF. It was reported to be less biased than the Gini VIM in simulation studies (Nicodemus and Malley, 2009; Strobl *et al.*, 2007b), and has been applied in several empirical studies (Diaz-Uriarte and Alvarez de Andres, 2006; Geurts *et al.*, 2005; Shen *et al.*, 2007). RF calculates two misclassification rates (MCR) for each predictor in a single fitted tree. One (MCR1) is from prediction of the outcome using

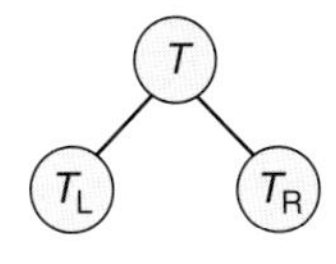

Figure 4.5. Node splitting of a binary decision tree.

each OOB sample. The other (MCR2) is from prediction using the permutated OOB samples (scrambling the values of the variable of interest). The permutation VIM is defined as the averaged increase (ΔMCR=MCR1−MCR2) between the two misclassification rates over all fitted trees in the RF. Assuming a normal distribution of ΔMCR, a *p*-value can be calculated from a *z*-score of the point estimate (E/SE). The randomly scrambling of the predictor X_i breaks the original association between X_i and outcome *Y*. Thus, MCR2 represents the null distribution of MCR when no association is observed. When there is a true association between X_i and *Y*, MCR1 should be substantially less than MCR2. Thus, the increase of MCR (ΔMCR) measures the importance of X_i in predicting the outcome *Y*.

The Gini-based VIM is computationally efficient but is prone to be biased under certain conditions. Two types of biases of VIM have been evaluated for RF to help the investigator to properly design and implement their analytical strategy. In a simulation study, Strobl *et al.* investigated the impact of data types on VIM (Strobl *et al.*, 2007a,b). The Gini-based CART is known to favor predictors with more categories in variable selection (Breiman *et al.*, 1984; Kim and Loh, 2001; Strobl *et al.*, 2007a). An alternative RF function, cforest, was developed based on unbiased classification trees (Hothorn *et al.*, 2006) without using the Gini index for tree growing. In the null case where none of the predictor variables are informative for the outcome, the VIM should not prefer any one type of predictor (e.g., continuous) over another (e.g., binary). In this scenario, both Gini and permutation VIM of RF biased toward more frequent selection of predictors with more categories, which was consistent with the observations of CART. This bias of variable selection diminished by combining the cforest with bootstrap sampling without replacement. Using the permutation VIM of RF, the means of the importance were not affected by the number of categories of a predictor. However, the variation of the permutation VIM was higher for the predictors with mor categories. Combining permutation VIM of cforest with bootstrap sampling without replacement, this inflation of variance for predictors with more categories also diminished. In the second scenario of simulation, an informative binary predictor was simulated with four uninformative predictors, one normally distributed continuous predictor and three categorical variables with various numbers of categories. Again, cforest with bootstrap sampling provided the most reliable selection of the informative predictor. The permutation VIM in both RF and cforest provided the most robust measurements to distinguish the informative and noninformative predictors. When the research question is focused on ranking predictors all in one type, such as transcriptomic predictors (continuous) or GWAS data (mostly with three genotypes), such bias caused by the mixture of multiple data types is unlikely to change the relative ranking of the predictors. However, this bias has to be considered and corrected when other types of predictors are analyzed jointly. For example, to include the gene-by-environment interactions (related discussions in III.B and IV.C) into the prediction model, the environmental factors which can be continuous and categorical (with various levels) need to be analyzed together with the genotypic data. In this situation, using VIM of cforest and bootstrap sampling without replacement is the most favorable approach to obtain unbiased assessment among both genetic and environmental predictors.

The second type of bias of variable selection in RF is introduced by the correlation between the predictors. Transcriptomic measurements can be correlated due to coregulation by common transcriptional factor or pathway. Genetic variants can also be highly correlated due to history of natural selection, gene mutation, and other evolutionary forces. The common genetics variants of human populations have been identified and measured in the HapMap projects (International HapMap Consortium, 2005; International HapMap Consortium *et al.*, 2007; The International HapMap Consortium, 2003). Most of these common variants are correlated in the form of common haplotypes. In population genetics, LD is the nonrandom association of alleles at two or more loci, regardless of their chromosomal location. LD describes a situation in which some combinations of alleles or genetic markers occur more or less frequently in a population than would be expected from a random formation of haplotypes from alleles based on their frequencies. The degree of LD measures the correlation between genetic variants at different loci.

Several studies reported that the correlation structure of predictors may cause bias in selecting the important predictors in RF. When the noise predictors (e.g., gene expression levels) were correlated, RF overestimated the importance in gene expression microarray analysis (Strobl *et al.*, 2008). Under the unconditional permutation scheme, RF highly overestimated the correlated predictors, particularly for small value of mtry (the number of variables used for each node splitting). A new, conditional permutation scheme was developed based on partitions generated by conditional inference forests (CIF) which provided unbiased VIM for different types of predictors (Hothorn *et al.*, 2006). The values of predictors were permuted for each tree by the partition of the space only by that tree. The VIM generated from the conditional permutation scheme greatly decreased the bias caused by correlated predictors (Strobl *et al.*, 2008).

One of the key questions for SNP analysis is that whether machine learning method like RF can distinguish the causal SNPs from the noncausal SNPs when strong correlation structure is present between them (i.e., high LD). In the case of RF, the distinction between the causal and noncausal SNPs is defined as more frequent selection by VIM. Meng *et al.* found that the power of identifying true risk SNPs could increase by strategically revising RF procedures of tree building and importance measurement calculation when SNPs were correlated measured by LD (Meng *et al.*, 2009). Gini VIM of RF was consistently found to be biased under correlation among predictors. In the case of selecting the genetic predictors from high-density GWAS data, the strong correlations among these genetic predictors can have a substantial impact on the behavior of the VIMs in RF. More importantly, a biased VIM may decrease the ability of RF in discriminating the correlated causal and noncausal genetic predictors. Based on their simulation study, Nicodemus and Malley (2009) recommended the following three analytical tips to ensure obtaining reliable variable selection results in high-dimensional "omic" data using RF. First, the number of trees grown by RF needs to be large. Second, the size of the terminal node needs to be large enough to avoid overfitting and to reduce the impact of strong correlation between predictors on variable selection. Third, RF needs to be run repeatedly to obtain a distribution of VIM (both point estimate and the variability) rather than to report and compare VIM from a single run of RF because the values of VIM obtained using permutation VIM varied from time to time. This will increase the computational burden but provide unbiased estimates of VIM. They also confirmed that CIF produced unbiased VIM and were not strongly affected by correlation between predictors. Therefore, CIF was preferred when studying correlated high-dimensional data such as SNP data from GWAS.

To further illustrate the potential biases of VIM of the original RF and CIF among correlated predictors, Nicodemus *et al.* (2010) extended the simulation studies of correlated predictors (Strobl *et al.*, 2008). To answer the question whether the original RF permutation VIM may be preferable to the conditional permutation VIM when strong correlation structure among predictors is present, intensive simulations were conducted for both H_0 (no predictors are associated with the outcome) and H_A (predictors are associated with the outcome in presence of correlation between predictors). The results suggested that conditional permutation VIM may be preferable in small-scale studies (the number of predictors is relatively small) where the goal is to identify causal predictors among a set of correlated ones. For large-scale studies, such as GWAS, the original permutation VIM of RF may be preferred to identify regions containing causal predictors.

RF is a nonlinear and nonparametric method and may not be expected to reproduce the same inferences as drawn from regression-based models. For variable selection purposes, the advantage of the RF VIM comparing to univariate screening methods is that RF covers the impact of each predictor individually (main effect) as well as in multivariate interactions with other predictors. Therefore, using VIM of RF may help identify additional biomarkers, which are associated with the outcome through a more complex, multiway interaction mechanism. These interaction effects, both gene–gene interactions and gene–environment interactions are very appealing to biologists, epidemiologists, and clinicians since they are closer to the real biological pathways, networks, and systems underlying the human diseases. These advantages of RF in the variable selection over the classic regression-based methods are further discussed in the following section.

C. Identification of interaction effects

Epistasis has been used as a general term to describe the complex interactions among genetic loci (Phillips, 2008). Three distinct meanings can be inferred by the term of "epistasis," namely "functional epistasis," "compositional epistatisis," and "statistical epistasis" (Phillips, 2008). The translation of the statistical interactions to the functional interactions has not been very successful and has been extensively debated (Cordell, 2009). The common genetic alleles identified by the GWAS using an additive genetic model without considering the epistatic associations can only explain a small fraction of the heritability. Because of the genetic factors commonly function through a biological system, it has been speculated that epistatic association is responsible for a part of the unexplained heritability (Manolio *et al.*, 2009; Moore and Williams, 2009).

Identification of epistatic effects among hundreds of thousands of SNPs posed a great computational and analytical challenge. Exhaustive test of high-order (e.g., more than three) interactions among genome-wide SNP data is practically not feasible without access to supercomputing facility. The huge number of tests also requires larger sample size to detect the true association by controlling for multiple testing. The traditional methods used in the genetic analysis (e.g., linear and logistic regression) are not capable of modeling the nonlinear effects among the biological processes. Thus, the machine learning methods are appealing to deal with the high-dimensional data and potentially nonlinear effects in the GWAS. RF, along with other machine learning methods, was proposed to detect a broader definition of "gene–gene interaction," which is distinct from the "statistical epistasis" described above. For these tree-based methods, a path in a tree is regarded as a form of interaction among linked predictors. RF potentially model large number of main effects as well as interaction effects in a computationally efficient approach without exhaustive searching for all possible combination of effects. Again, the inclusion of individual predictor and the interactions is decided by their contribution of fitting the data and improving the predictive ability of the outcome. Because multiple independent trees are fitted in a RF model, the effect of a specific interaction cannot be explicitly evaluated. The major contributions of RF in studying gene–gene interactions are (1) improving the predictive ability of outcome by including interaction effects, (2) identifying the predictors potentially involved in complex interactions rather than only their marginal effects. Recent development has improved the ability of detecting interaction effects using ensemble learning approach and is further discussed in Section V.

Whether including the gene–gene interactions improves prediction of the outcome is a largely unanswered question in complex disease. Although machine learning methods, such as RF, are unable to identify a specific interaction effect, they can serve as a useful tool to estimate whether we can better classify complex diseases, or how much we can improve the predictive ability by including the gene–gene interactions. Since the identified marginal effects of common genetic variants only explained a small portion of the heritability, the first hypothesis RF can address is H_0: No improvement of the predictive ability of disease Y by including the interaction effects of common genetic variants. By testing this hypothesis, we are able to assess the role of gene–gene interactions in complex disease, and to evaluate whether gene–gene interaction is a major component to explain the "missing heritability." In a recent genetic association study of leukoaraiosis volume measured by resonance imaging (MRI), the main effects, gene–gene interactions, and gene–environment interactions between these SNPs and covariates were examined (Smith *et al.*, 2009). In a model including additive SNP main effects, the four significant, replicated, and cross-validated SNPs joint explained 5.99% of variation in leukoaraiosis. A model consisting of the top four SNP–SNP interactions explained 14.73% of variation in leukoaraiosis. These results strongly indicated that combining the gene–gene interaction effects in the model may improve the prediction of the complex disease traits comparing to the model only consisting the main effects of genes.

To evaluate the second utility of RF in studying gene–gene interaction (identifying the predictors potentially involved in complex interactions rather than only their marginal effects), Lunetta *et al.* conducted a simulation study (Lunetta *et al.*, 2004) of 500 cases and 500 controls (100 replicates). Both risk SNPs (associated with the outcome) and noise SNPs (independent of the outcome) were simulated. When risk SNPs interacted and other factors were same, the RF VIM outperforms the univarite Fisher Exact test (for SNP main effect) as a variable selection tool. As the number of interacting SNPs increased, the improvement in performance of RF over Fisher Exact test for variable selection also increased. When no SNP interaction was simulated, the performance of RF and Fisher Exact test were essentially same for the purpose of variable selection. This study strongly demonstrated that RF as a variable selection tool can identify truly associated SNPs with either main effect or interaction effect in high-dimensional genetic association studies.

Tang *et al.* (2009) developed a permutation-based approach to evaluate the gene–gene interaction using RF. The genotypes at all SNPs within a gene of interest among the subjects were permuted while the LD structure within that gene was maintained. For a SNP not in the permuted gene, a decrease in VIM may indicate a gene–gene interaction between the SNP and the permuted gene. This approach of combining the variable selection function of RF with permutation procedure was applied in analyzing rheumatoid arthritis at GAW16 but no strong evidence of gene–gene interaction was identified.

Similarly, the multiway interaction between genetic risk factors and environmental risk factors can also be modeled using RF for predicting the outcome. Because there are usually a small number of environmental risk factors comparing to the genetic risk factors, the computational efficiency of RF is less appealing when identifying the gene–environment interaction is the primary goal, especially for identifying pair-wise gene–environment interaction. For a classification problem involving both genetic and environmental risk factors, it is still a valid research question to ask which genetic factors are important in the context of a set of environmental factors. By answering this question, we will be able to better understand those genetic factors functioning in the context-dependent way. This is also an essential component of the concept of genetic architecture. RF provides a nonparametric framework to model and rank both genetic and environmental factors by measuring their importance in predicting the outcome. In a study of predicting CAC, a marker of subclinical coronary atherosclerosis, 471 SNPs on candidate gene along with 17 other risk factors were compared by their VIM using RF (Sun *et al.*, 2008). Among the top 50 predictors, the same eight SNPs and 12 risk factors (age, body mass index, sex, serum glucose, high-density lipoprotein cholesterol, systolic blood pressure, cholesterol, homocysteine, triglycerides, fibrinogen, Lp(a) and low-density lipoprotein particle size) were found in two replicate datasets sampled from sibships. These SNPs were selected likely through the interactions with the risk factors since the main effects of these SNPs were not significantly associated with CAC (unpublished data). As discussed in Section IV.B, sampling strategy and VIM may introduce bias in variable selection. Because SNPs are categorical variables with up to three categories, they are less favored to be selected by the original RF comparing to the continuous risk factors or risk factors with more categories. In order to rank the genetic and environmental factors without bias, combining cforest with bootstrap sampling without replacement is a preferred alternative method (Strobl *et al.*, 2007b).

V. INTERPRETATION OF THE RESULTS

Many of the advanced machine learning methods, such as neural networks and SVM, are so-called "black box" learners. They are capable of classifying the outcome by creating artificial features while fitting the training data. However, the role of individual predictor in predicting the outcome can hardly be estimated. The VIMs in RF and other related methods provide a way to uncover the relative importance of the predictors. In studies of genetic factors, consistent high ranking of the genetic predictors suggests the plausible biological relationship with the disease outcome (Sun *et al.*, 2008). Since VIMs are calculated in a flexible tree-based computation framework, this feature of RF in genetic studies has to be cautiously interpreted with other statistical and biological evidence (Sun *et al.*, 2008). The interpretability of a RF is not as straightforward as that of an individual classification tree, where the influence of a predictor variable directly corresponds to its position in the tree. Incorporating the interactions into RF model for prediction and variable selection is an advantage for improving the performance of prediction, but it is also a disadvantage for poor interpretability since none of the interactions in the heterogeneous trees can be explicitly identified and related to the outcome.

RuleFit is a newly developed ensemble learning method which inherits the same advantages and adds more interpretability to the model (Friedman and Popescu, 2005). RuleFit is based on the ensembles of rules and each rule consists of a series of simple statements concerning the independent variables. As indicated in the simulation study, these rule ensembles produce predictive accuracy comparable to the best methods in the field. Because of its simple form, each rule is easy to understand and to interpret the underlying relationships among variables. Moreover, each rule's influence on the predictive model and the relative importance of each independent variable can be assessed by the algorithm. For high-dimensional phenotypic and genotypic data, the RuleFit would be able to provide: (1) an accurate predictive model for the target outcome. (2) A set of simple rules which could reflect gene–gene interactions as well as gene–environment interactions. (3) A variable selection method by ranking the relative importance of all independent variables.

RF and other tree-based ensemble learning machines are nonlinear and nonparametric methods. They are not expected to reproduce inferences drawn from regression-based statistical models. These machine learning methods cannot replace the regression-based models when explicit interpretation of the relationship between the predictor and outcome is the expectation. However, when accurate prediction and effective screening are the goals, RF is a strong candidate to achieve these goals for high-dimensional genomic data and provides a semi-interpretable measurement of predictors to follow up.

Inferring molecular function cannot be achieved by statistical and computational methods alone. However, the evidence of putative molecular function can be enriched by accumulating multilayer information using other bioinformatic tools. For example, a nonsynonymous SNP causing amino acid change is regarded as putative functional. When RF identifies such a SNP (rs3749172) in GPR35 (Sun *et al.*, 2008) linked to CAC, a follow-up sequence analysis of GPR35 protein revealed a potential functional change specific to this protein caused by this nonsynonymous polymorphism (Fig. 4.6). GPR35 encodes a highly conserved (in human, mouse, and rat) G-protein-coupled receptor with seven transmembrane domains (1–7 in Fig. 4.6). The C-terminal of GPR35 protein binds G protein to conduct the extracellular signal into the cytosol. The phosphorylation of the C-terminal is essential for the function of this signal transduction process. Four conserved amino acids (labeled by the arrows in Fig. 4.6), Serine or Threonine (with a hydroxyl group), are the potential sites for phosphorylation. SNP rs3749172 (A/C polymorphism) corresponds to a missense change from reference Serine (A allele) to Arginine (C allele) at

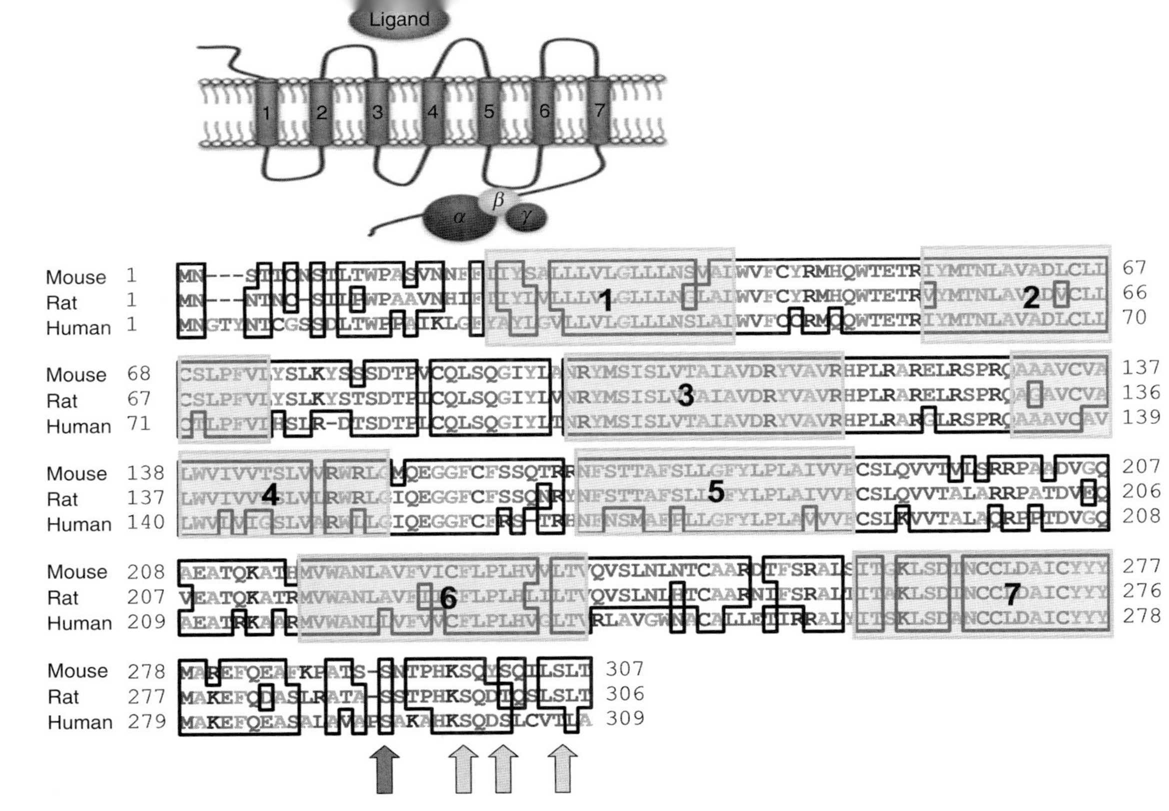

Figure 4.6. Sequence analysis of the conservation of GPR35 protein of human, mouse, and rat. The upper figure illustrates the domains of a GPR. The N-terminal extracellular domain recognizes the ligand containing the external signal and the C-terminal cytosolic domain binds the G protein. The lower figure represents the sequence alignment of GPR35 proteins of mouse, rat, and human. The seven transmembrane domains are conserved and labeled 1–7. Four conserved amino acids at the C-terminal are plausible sites for phosphorylation. Among the four marked amino acids (by arrows), the human Ser294 (by dark arrow) corresponds to the nonsynonymous rs3749172.

amino acid 294 of human GPR35 (labeled by a dark arrow in Fig. 4.6). This missense change not only causes a dramatic shift of side chain chemistry from a neutral amino acid (Serine) to a basic amino acid (Arginine), but also eliminates one of the four conserved phosphorylation sites that are essential for the protein's core function. Analyzing the impact of genetic variants on gene expression levels can also provide evidence for further support of their functional roles. Databases of associations between genetic variants and global gene expression levels are available in multiple human tissues and cell lines (Dixon *et al.*, 2007; Goring *et al.*, 2007; Schadt *et al.*, 2008). Searching these databases directly (Kottgen *et al.*, 2010) or using the online bioinformatic tool such as SCAN (Gamazon *et al.*, 2010) provides additional confidence of the functionality of an identified genetic variant. Having these bioinformatic and biochemical supports along with the analytical evidence from RF, a more confidant conclusion of the importance of the genetic factor can be made, and a more specific hypothesis can be generated to follow up.

VI. CONCLUSIONS

The analysis always needs to be consistent with the research question and study design to obtain robust results, and to reach the final conclusion. RF is a powerful analytical tool, but its utility can only be demonstrated when all these rerated aspects of a study to be cautiously considered. RF can assist to answer primarily two classes of research questions, predicting the outcome and selecting predictive variables, in the context of genetic association study and GWAS. RF was originally developed as a prediction tool. It was known to be capable of handling high-dimensional data and to be resistant to uninformative predictors. The high dimensionality is a relative definition. For machine learning, high dimensionality can mean $p \approx n$ or $p \gg n$, where p is the number of prediction variables and n is the number of samples. There is no report about the predictive ability of RF in very high-dimensional data, such as GWAS, where p is usually hundred folds higher than n. Under such scenario, how to achieve good performance of RF for prediction and whether RF is still computationally efficient are unanswered questions. In GWAS data, most of the measured genetic variants are not expected to be related to the outcome of interest. Therefore, the number of uninformative variables is extremely high. In this case, when the overall signal-to-noise ratio is too low, the predictive ability of RF can decrease as discussed in Section IV.A. In order to maximize the utility of RF in prediction, a screening step to enrich the informative markers and to reduce the dimensionality is a reasonable consideration. The goal of variable selection can be achieved by RF itself, and other statistical methods such as regularized regression methods. The advantage of RF in variable selection lies in its capability of capturing a large number of gene–gene and gene–environment interactions in addition to the main effects of these variables (Section IV.B). When these interaction effects are the major component of the outcome, RF is a more appropriate choice than the main effect models. A two-step RF design, variable selection using RF followed by model fitting using RF, may also help to build an accurate prediction model of a complex trait using GWAS data. So far, the utilities of RF have only been studied in simulation studies and lower dimensional genetic data analysis. With the newly implemented RF application (www.randomjungle.org) which addressed the computational efficiency, and other modifications to address the limitations of original RF, RF may eventually demonstrate its utility of analyzing GWAS data and mapping the complex genetic architecture of human diseases.

References

Altshuler, D., Daly, M. J., and Lander, E. S. (2008). Genetic mapping in human disease. *Science* **322,** 881–888.

Amaratunga, D., Cabrera, J., and Lee, Y. S. (2008). Enriched random forests. *Bioinformatics* **24,** 2010–2014.

Aulchenko, Y. S., Ripke, S., Isaacs, A., and van Duijn, C. M. (2007). GenABEL: An R library for genome-wide association analysis. *Bioinformatics* **23,** 1294–1296.

Breiman, L. (2001). Random forests. *Mach. Learn.* **45,** 5–32.

Breiman, L., Friedman, J. H., Olshen, R. A., and Stone, C. J. (1984). Classification and Regression Trees. Chapman & Hall, New York.

Bureau, A., Dupuis, J., Falls, K., Lunetta, K. L., Hayward, B., Keith, T. P., and Van Eerdewegh, P. (2005). Identifying SNPs predictive of phenotype using random forests. *Genet. Epidemiol.* **28,** 171–182.

Chen, X. W., and Liu, M. (2005). Prediction of protein–protein interactions using random decision forest framework. *Bioinformatics* **21,** 4394–4400.

Christensen, B. C., Houseman, E. A., Godleski, J. J., Marsit, C. J., Longacker, J. L., Roelofs, C. R., Karagas, M. R., Wrensch, M. R., Yeh, R. F., Nelson, H. H., *et al.* (2009). Epigenetic profiles distinguish pleural mesothelioma from normal pleura and predict lung asbestos burden and clinical outcome. *Cancer Res.* **69,** 227–234.

Cordell, H. J. (2009). Genome-wide association studies: Detecting gene–gene interactions that underlie human diseases. *Nat. Rev. Genet.* **10,** 392–404.

Diaz-Uriarte, R., and Alvarez de Andres, S. (2006). Gene selection and classification of microarray data using random forest. *BMC Bioinformatics* **7,** 3.

Dixon, A. L., Liang, L., Moffatt, M. F., Chen, W., Heath, S., Wong, K. C., Taylor, J., Burnett, E., Gut, I., Farrall, M., *et al.* (2007). A genome-wide association study of global gene expression. *Nat. Genet.* **39,** 1202–1207.

Flint, J., and Mackay, T. F. (2009). Genetic architecture of quantitative traits in mice, flies, and humans. *Genome Res.* **19,** 723–733.

Frazer, K. A., Murray, S. S., Schork, N. J., and Topol, E. J. (2009). Human genetic variation and its contribution to complex traits. *Nat. Rev. Genet.* **10,** 241–251.

Friedman, J. H. and Popescu, B. E. (2005). Predictive learning via rule ensembles. Technical Report. Stanford University.

Gamazon, E. R., Zhang, W., Konkashbaev, A., Duan, S., Kistner, E. O., Nicolae, D. L., Dolan, M. E., and Cox, N. J. (2010). SCAN: SNP and copy number annotation. *Bioinformatics* **26,** 259–262.

Geurts, P., Fillet, M., de Seny, D., Meuwis, M. A., Malaise, M., Merville, M. P., and Wehenkel, L. (2005). Proteomic mass spectra classification using decision tree based ensemble methods. *Bioinformatics* **21,** 3138–3145.

Goring, H. H., Curran, J. E., Johnson, M. P., Dyer, T. D., Charlesworth, J., Cole, S. A., Jowett, J. B., Abraham, L. J., Rainwater, D. L., Comuzzie, A. G., *et al.* (2007). Discovery of expression QTLs using large-scale transcriptional profiling in human lymphocytes. *Nat. Genet.* **39,** 1208–1216.

Guyton, A. C., Coleman, T. G., and Granger, H. J. (1972). Circulation: Overall regulation. *Annu. Rev. Physiol.* **34,** 13–46.

Hansen, L. K., and Salomon, P. (1990). Neural network ensembles. *IEEE Trans. Pattern Anal. Mach. Intell.* **12,** 993–1001.

Hastie, T., Tibshirani, R., and Friedman, J. (2001). The Elements of Statistical Learning: Data Mining, Inference, and Prediction. Springer, New York.

Hindorff, L. A., Sethupathy, P., Junkins, H. A., Ramos, E. M., Mehta, J. P., Collins, F. S., and Manolio, T. A. (2009). Potential etiologic and functional implications of genome-wide association loci for human diseases and traits. *Proc. Natl. Acad. Sci. USA* **106,** 9362–9367.

Hothorn, T., Hornik, K., and Zeileis, A. (2006). Unbiased recursive partitioning. *J. Comput. Graph. Stat.* **15,** 651–674.

Huang, X., Pan, W., Grindle, S., Han, X., Chen, Y., Park, S. J., Miller, L. W., and Hall, J. (2005). A comparative study of discriminating human heart failure etiology using gene expression profiles. *BMC Bioinformatics* **6,** 205.

International HapMap Consortium (2005). A haplotype map of the human genome. *Nature* **437,** 1299–1320.

International HapMap Consortium Frazer, K. A., Ballinger, D. G., Cox, D. R., Hinds, D. A., Stuve, L. L., Gibbs, R. A., Belmont, J. W., Boudreau, A., Hardenbol, P., *et al.* (2007). A second generation human haplotype map of over 3.1 million SNPs. *Nature* **449,** 851–861.

Izmirlian, G. (2004). Application of the random forest classification algorithm to a SELDI-TOF proteomics study in the setting of a cancer prevention trial. *Ann. N.Y. Acad. Sci.* **1020,** 154–174.

Kim, H., and Loh, W. Y. (2001). Classification trees with unbiased multiway splits. *J. Am. Stat. Assoc.* **96,** 589–604.

Kottgen, A., Pattaro, C., Boger, C. A., Fuchsberger, C., Olden, M., Glazer, N. L., Parsa, A., Gao, X., Yang, Q., Smith, A. V., *et al.* (2010). New loci associated with kidney function and chronic kidney disease. *Nat. Genet.* **42,** 376–384.

Kuncheva, L. I., and Whitaker, C. J. (2003). Measures of diversity in classifier ensembles and their relationship with the ensemble accuracy. *Mach. Learn.* **51,** 181–207.

Lunetta, K. L., Hayward, L. B., Segal, J., and Van Eerdewegh, P. (2004). Screening large-scale association study data: Exploiting interactions using random forests. *BMC Genet.* **5,** 32.

Manolio, T. A., Collins, F. S., Cox, N. J., Goldstein, D. B., Hindorff, L. A., Hunter, D. J., McCarthy, M. I., Ramos, E. M., Cardon, L. R., Chakravarti, A., *et al.* (2009). Finding the missing heritability of complex diseases. *Nature* **461,** 747–753.

Meng, Y. A., Yu, Y., Cupples, L. A., Farrer, L. A., and Lunetta, K. L. (2009). Performance of random forest when SNPs are in linkage disequilibrium. *BMC Bioinformatics* **10,** 78.

Moore, J. H., and Williams, S. M. (2009). Epistasis and its implications for personal genetics. *Am. J. Hum. Genet.* **85,** 309–320.

Moore, J. H., Asselbergs, F. W., and Williams, S. M. (2010). Bioinformatics challenges for genome-wide association studies. *Bioinformatics* **26,** 445–455.

Morrison, A. C., Bare, L. A., Chambless, L. E., Ellis, S. G., Malloy, M., Kane, J. P., Pankow, J. S., Devlin, J. J., Willerson, J. T., and Boerwinkle, E. (2007). Prediction of coronary heart disease risk using a genetic risk score: The Atherosclerosis Risk in Communities Study. *Am. J. Epidemiol.* **166,** 28–35.

Nicodemus, K. K., and Malley, J. D. (2009). Predictor correlation impacts machine learning algorithms: Implications for genomic studies. *Bioinformatics* **25,** 1884–1890.

Nicodemus, K. K., Malley, J. D., Strobl, C., and Ziegler, A. (2010). The behaviour of random forest permutation-based variable importance measures under predictor correlation. *BMC Bioinformatics* **11,** 110.

Paynter, N. P., Chasman, D. I., Pare, G., Buring, J. E., Cook, N. R., Miletich, J. P., and Ridker, P. M. (2010). Association between a literature-based genetic risk score and cardiovascular events in women. *JAMA* **303,** 631–637.

Phillips, P. C. (2008). Epistasis—The essential role of gene interactions in the structure and evolution of genetic systems. *Nat. Rev. Genet.* **9,** 855–867.

Purcell, S., Neale, B., Todd-Brown, K., Thomas, L., Ferreira, M. A., Bender, D., Maller, J., Sklar, P., de Bakker, P. I., Daly, M. J., *et al.* (2007). PLINK: A tool set for whole-genome association and population-based linkage analyses. *Am. J. Hum. Genet.* **81,** 559–575.

Qi, Y., Klein-Seetharaman, J., and Bar-Joseph, Z. (2005). Random forest similarity for protein–protein interaction prediction from multiple sources. *Pac. Symp. Biocomput.* 531–542.

Savoia, C., and Schiffrin, E. L. (2006). Inflammation in hypertension. *Curr. Opin. Nephrol. Hypertens.* **15,** 152–158.

Schadt, E. E., Molony, C., Chudin, E., Hao, K., Yang, X., Lum, P. Y., Kasarskis, A., Zhang, B., Wang, S., Suver, C., *et al.* (2008). Mapping the genetic architecture of gene expression in human liver. *PLoS Biol.* **6,** e107.

Shen, K. Q., Ong, C. J., Li, X. P., Hui, Z., and Wilder-Smith, E. (2007). A feature selection method for multilevel mental fatigue EEG classification. *IEEE Trans. Biomed. Eng.* **54,** 1231–1237.

Shi, T., Seligson, D., Belldegrun, A. S., Palotie, A., and Horvath, S. (2005). Tumor classification by tissue microarray profiling: Random forest clustering applied to renal cell carcinoma. *Mod. Pathol.* **18,** 547–557.

Sing, C. F., and Boerwinkle, E. A. (1987). Genetic architecture of inter-individual variability in apolipoprotein, lipoprotein and lipid phenotypes. *Ciba Found. Symp.* **130,** 99–127.

Smith, J. A., Turner, S. T., Sun, Y. V., Fornage, M., Kelly, R. J., Mosley, T. H., Jack, C. R., Kullo, I. J., and Kardia, S. L. (2009). Complexity in the genetic architecture of leukoaraiosis in hypertensive sibships from the GENOA Study. *BMC Med. Genomics* **2,** 16.

Strobl, C., Boulesteix, A. L., and Augustin, T. (2007a). Unbiased split selection for classification trees based on the Gini Index. *Comput. Stat. Data Anal.* **52,** 483–501.

Strobl, C., Boulesteix, A. L., Zeileis, A., and Hothorn, T. (2007b). Bias in random forest variable importance measures: Illustrations, sources and a solution. *BMC Bioinformatics* **8,** 25.

Strobl, C., Boulesteix, A. L., Kneib, T., Augustin, T., and Zeileis, A. (2008). Conditional variable importance for random forests. *BMC Bioinformatics* **9,** 307.

Sun, Y. V., Cai, Z., Desai, K., Lawrance, R., Leff, R., Jawaid, A., Kardia, S. L., and Yang, H. (2007). Classification of rheumatoid arthritis status with candidate gene and genome-wide single-nucleotide polymorphisms using random forests. *BMC Proc.* **1**(Suppl. 1), S62.

Sun, Y. V., Bielak, L. F., Peyser, P. A., Turner, S. T., Sheedy, P. F., II, Boerwinkle, E., and Kardia, S. L. (2008). Application of machine learning algorithms to predict coronary artery calcification with a sibship-based design. *Genet. Epidemiol.* **32,** 350–360.

Talmud, P. J., Hingorani, A. D., Cooper, J. A., Marmot, M. G., Brunner, E. J., Kumari, M., Kivimaki, M., and Humphries, S. E. (2010). Utility of genetic and non-genetic risk factors in prediction of type 2 diabetes: Whitehall II prospective cohort study. *BMJ* **340,** b4838.

Tang, R., Sinnwell, J. P., Li, J., Rider, D. N., de Andrade, M., and Biernacka, J. M. (2009). Identification of genes and haplotypes that predict rheumatoid arthritis using random forests. *BMC Proc.* **3**(Suppl. 7), S68.

The International HapMap Consortium (2003). The International HapMap Project. *Nature* **426,** 789–796.

Wu, B., Abbott, T., Fishman, D., McMurray, W., Mor, G., Stone, K., Ward, D., Williams, K., and Zhao, H. (2003). Comparison of statistical methods for classification of ovarian cancer using mass spectrometry data. *Bioinformatics* **19,** 1636–1643.

5 Detecting, Characterizing, and Interpreting Nonlinear Gene–Gene Interactions Using Multifactor Dimensionality Reduction

Jason H. Moore
Institute for Quantitative Biomedical Sciences, Departments of Genetics and Community and Family Medicine, Dartmouth Medical School, Lebanon, New Hampshire, USA

ABSTRACT

Human health is a complex process that is dependent on many genes, many environmental factors and chance events that are perhaps not measurable with current technology or are simply unknowable. Success in the design and execution of population-based association studies to identify those genetic and environmental factors that play an important role in human disease will depend on our ability to embrace, rather that ignore, complexity in the genotype to

Advances in Genetics, Vol. 72

0065-2660/10 $35.00
DOI: 10.1016/S0065-2660(10)72005-8

phenotype mapping relationship for any given human ecology. We review here three general computational challenges that must be addressed. First, data mining and machine learning methods are needed to model nonlinear interactions between multiple genetic and environmental factors. Second, filter and wrapper methods are needed to identify attribute interactions in large and complex solution landscapes. Third, visualization methods are needed to help interpret computational models and results. We provide here an overview of the multifactor dimensionality reduction (MDR) method that was developed for addressing each of these challenges.

I. INTRODUCTION

Human genetics has a long and rich history of research to understand the role of interindividual variation in the human genome and variation in biological traits. We have progress rapidly from unmeasured genetic studies in families to the identification of common variation in the DNA sequence that can be used in population-based association studies. This is an exciting time because we now have access to technology that allows us to efficiently measure many DNA sequence variations from across the human genome. We will within the next 5 years likely have access to cutting-edge technology that will deliver the entire genomic sequence for all subjects in our genetic and epidemiologic studies. Now that we have access to the basic hereditary information it is time to shift our focus toward the analysis of this data. The focus of this chapter is on the important role of computer science, and, more specifically, machine learning for mining patterns of genetic variations that are associated with susceptibility to common human diseases. This approach assumes that the relationship between genotype and phenotype is very complex. Specifically, we will focus on computational methods for identifying gene–gene interactions or epistasis that accounts for part of the complexity of genetic architecture.

A. The role of epistasis in common human diseases

Human genetics has been largely successful in identifying the causative mutations in single genes that determine with virtual certainly rare diseases such as sickle-cell anemia. However, the same success has not been had for common human diseases such as sporadic breast cancer, essential hypertension or bipolar depression. This is because diseases that are common in the population have a much more complex etiology that requires different research strategies than were used to identify genes underlying rare diseases that follow a simpler Mendelian inheritance pattern. Complexity can arise from phenomena such as locus heterogeneity (i.e., different DNA sequence variations leading to the same phenotype), phenocopy (i.e., environmentally determined phenotypes), and the dependence of genotypic effects on environmental factors (i.e., gene–environment interactions or plastic reaction norms) and genotypes at other loci (i.e., gene–gene interactions or epistasis). It is this latter source of complexity, epistasis, that is of interest here. Epistasis has been recognized for many years as deviations from the simple inheritance patterns observed by Mendel (Bateson, 1909) or deviations from additivity in a linear statistical model (Fisher, 1918) and is likely due, in part, to canalization or mechanisms of stabilizing selection that evolve robust (i.e., redundant) gene networks (Waddington, 1942).

Epistasis has been defined in multiple different ways (e.g., Phillips, 1998, 2008). We have reviewed two types of epistasis, biological and statistical (Moore and Williams, 2005, 2009; Tyler *et al.*, 2009). Biological epistasis when the physical interactions between biomolecules (e.g., DNA, RNA, proteins, enzymes, etc.) are influenced by genetic variation at multiple different loci. This type of epistasis occurs at the cellular level in an individual and is what Bateson (1909) had in mind when he coined the term. Statistical epistasis, on the other hand, occurs at the population level and is realized when there is interindividual variation in DNA sequences. The statistical phenomenon of epistasis is what Fisher (1918) had in mind. The relationship between biological and statistical epistasis is often confusing but will be important to understand if we are to make biological inferences from statistical results (Moore and Williams, 2005, 2009; Phillips, 1998, 2008; Tyler *et al.*, 2009). Moore (2003) has argued that epistasis is likely to be a ubiquitous phenomenon in complex human diseases. The focus of the present study is the detection and characterization of statistical epistasis in human populations using machine learning and data mining methods.

B. Computational challenges

The fields of genetics and epidemiology are undergoing an information explosion and an understanding implosion. That is, our ability to generate data is far outpacing our ability to interpret it. This is especially true today where it is technically and economically feasible to measure a million or more single nucleotide polymorphisms (SNPs) from across the human genome. An important goal in human genetics is to determine which of the millions of SNPs are useful for predicting who is at risk for common diseases. This "genome-wide" approach is expected to revolutionize the genetic analysis of common human diseases and, for better or worse, is quickly replacing the traditional "candidate-gene" approach that focuses on several genes selected by their known or suspected function.

Moore and Ritchie (2004) have outlined three significant challenges that must be overcome if we are to successfully identify genetic predictors of health and disease using a genome-wide approach. First, powerful data mining and machine learning methods will need to be developed to statistically model the relationship between combinations of DNA sequence variations and disease susceptibility. Traditional methods such as logistic regression have limited power for modeling high-order nonlinear interactions (Moore and Williams, 2002). A second challenge is the selection of genetic features or attributes that should be included for analysis. If interactions between genes explain most of the heritability of common diseases, then combinations of DNA sequence variations will need to be evaluated from a list of thousands of candidates. Filter (SNP selection) and wrapper (SNP searching) methods will play an important role because there are more combinations than can be exhaustively evaluated. A third challenge is the interpretation of gene–gene interaction models. Although a statistical model can be used to identify DNA sequence variations that confer risk for disease, this approach cannot be translated into specific prevention and treatment strategies without interpreting the results in the context of human biology. Making etiological inferences from computational models may be the most important and the most difficult challenge of all (Moore and Williams, 2005).

To illustrate the concept of statistical interaction, consider the following simple example of epistasis in the form of a penetrance function. Penetrance is simply the probability (*P*) of disease (D) given a particular combination of genotypes (G) that was inherited (i.e., $P[D|G]$). Let us assume for two SNPs labeled A and *B* that genotypes *AA*, *aa*, *BB*, and *bb* have population frequencies of 0.25, while genotypes *Aa* and *Bb* have frequencies of 0.5. Let us also assume that individuals have a very high risk of disease if they inherit *Aa* or *Bb* but not both (i.e., the exclusive OR or XOR logic function). What makes this model interesting is that disease risk is entirely dependent on the particular *combination* of genotypes inherited at more than one locus. The penetrance for each individual genotype in this model is all the same and is computed by summing the products of the genotype frequencies and penetrance values. Heritability can be calculated as outlined by Culverhouse *et al.* (2002). Thus, in this model there is no difference in disease risk for each single-locus genotype as specified by penetrance values. This model is labeled M170 by Li and Reich (2000) in their categorization of genetic models involving two SNPs and is an example of a pattern that is not separable by a simple linear function. This model is a special case where all of the heritability is due to epistasis or nonlinear gene–gene interaction.

Combining this type of statistical interaction with the challenge of variable selection yields what computer scientists have called a *needle-in-a-haystack* problem. That is, there may be a particular combination of SNPs or

SNPs and environmental factors that together with the right nonlinear function are a significant predictor of disease susceptibility. However, individually they may not look any different than thousands of other SNPs that are not involved in the disease process and are thus noisy. Under these models, the learning algorithm is truly looking for a genetic needle in a genomic haystack. It is now commonly assumed that at least 1,000,000 carefully selected SNPs may be necessary to capture all of the relevant variation across the Caucasian human genome. Assuming this is true, we would need to scan approximately 500 billion pairwise combinations of SNPs to find a genetic needle. The number of higher order combinations is astronomical. What is the optimal computational approach to this problem?

There are two general approaches to select attributes for predictive models. The filter approach preprocesses the data by algorithmically, statistically, or biologically assessing the quality or relevance of each variable and then using that information to select a subset for analysis. The wrapper approach iteratively selects subsets of attributes for classification using either a deterministic or stochastic algorithm. The key difference between the two approaches is that the learning algorithm plays no role in selecting which attributes to consider in the filter approach. The advantage of the filter is speed while the wrapper approach has the potential to do a better job classifying subjects as sick or healthy. We first discuss a specific machine learning algorithm called multifactor dimensionality reduction (MDR) that has been applied to classifying healthy and disease subjects using their DNA sequence information and then discuss filter and wrapper approaches for the specific problem of detecting epistasis or gene–gene interactions on a genome-wide scale.

II. MACHINE LEARNING ANALYSIS OF GENE–GENE INTERACTIONS USING MDR

As discussed above, one of the early definitions of epistasis was deviation from additivity in a linear model (Fisher, 1918). The linear model plays a very important role in modern genetics and epidemiology because it has solid theoretical foundation, is easy to implement using a wide range of different software packages, and is easy to interpret. Despite these good reasons to use linear models, they do have limitations for detecting nonlinear patterns of interaction (e.g., Moore and Williams, 2002). The first problem is that modeling interactions require looking at combinations of attributes. Considering multiple attributes simultaneously is challenging because the available data get spread thinly across multiple combinations of genotypes, for example. Estimation of parameters in a linear model can be problematic when the data are sparse. The second problem is that linear models are often implemented such that interaction effects are only considered after independent main effects are identified. This certainly makes model fitting easier but assumes that the important predictors will have main effects. Further, it is well documented that linear models have greater power to detect main effects than interactions (e.g., Lewontin, 1974). For example, the FITF approach of Millstein *et al.* (2006) provides a powerful logistic regression approach to detecting interactions but conditions on marginal effects. Moore (2003) argues that this is an unrealistic assumption for common human diseases. The limitations of the linear model and other parametric statistical approaches have motivated the development of computational approaches such as those from machine learning and data mining that make fewer assumptions about the functional form of the model and the effects being modeled (McKinney *et al.*, 2006). As reviewed recently by Cordell (2009), MDR has emerged as one important new and novel method for detecting and characterizing patterns of statistical epistasis in genetic association studies that complements the linear modeling paradigm.

MDR was developed as a nonparametric (i.e., no parameters are estimated) and genetic model-free (i.e., no genetic model is assumed) machine learning strategy for identifying combinations of discrete genetic and environmental factors that are predictive of a discrete clinical endpoint (Hahn *et al.*, 2003; Moore, 2004, 2007; Moore *et al.*, 2006; Pattin *et al.*, 2009; Ritchie *et al.*, 2001, 2003; Velez *et al.*, 2007). Unlike most other methods, MDR was designed to detect interactions in the absence of detectable main effects and thus complements approaches such as logistic regression and random forests. At the heart of the MDR approach is a feature or attribute construction algorithm that creates a new variable or attribute by pooling, for example, genotypes from multiple SNPs. The process of defining a new attribute as a function of two or more other attributes is referred to as constructive induction or attribute construction and was first developed by Michalski (1983). Constructive induction using the MDR kernel is accomplished in the following way. Given a threshold T, a multilocus genotype combination is considered high risk if the ratio of cases (subjects with disease) to controls (healthy subjects) exceeds T, else it is considered low risk. Genotype combinations considered to be high risk are labeled G_1 while those considered low risk are labeled G_0. This process constructs a new 1D attribute with levels G_0 and G_1. It is this new single variable that is assessed using any classification method. The MDR method is based on the idea that changing the representation space of the data will make it easier for a classifier such as a decision tree or a naive Bayes learner to detect attribute dependencies. Cross-validation and permutation testing are used to prevent overfitting and false-positives due to multiple testing. Open-source software in Java and C are freely available from www.epistasis.org.

Since its initial description by Ritchie *et al.* (2001) many modifications and extensions to MDR have been proposed. These include, for example, entropy-based interpretation methods (Moore *et al.*, 2006), the use of odds ratios (Chung *et al.*, 2007) and Fisher's exact test (Gui *et al.*, 2010b) for more robust models, log-linear methods (Lee *et al.*, 2007), generalized linear models (Lou *et al.*, 2007), methods for imbalanced data (Velez *et al.*, 2007), model-based methods (Calle *et al.*, 2008, 2010), permutation testing methods (Edwards *et al.*, 2010; Greene *et al.*, 2010a; Pattin *et al.*, 2009), methods for missing data (Namkung *et al.*, 2009a), methods for covariate adjustment (Gui *et al.*, 2010b), methods for family data (Cattaert *et al.*, 2010; Lou *et al.*, 2008; Martin *et al.*, 2006) and different evaluation metrics (Bush *et al.*, 2008; Mei *et al.*, 2007; Namkung *et al.*, 2009b). The MDR approach has also been successfully applied to a wide range of different genetic association studies. For example, Andrew *et al.* (2006) used MDR to model the relationship between polymorphisms in DNA repair enzyme genes and susceptibility to bladder cancer. A highly significant nonadditive interaction was found between two SNPs in the *xeroderma pigmentosum group D* (*XPD*) gene that was a better predictor of bladder cancer than smoking. Interestingly, neither of these polymorphisms had a significant marginal effect. These interaction results were later replicated in independent studies from a consortium (Andrew *et al.*, 2008).

As discussed above, the biggest challenge to implementing methods such as MDR in on a genome-wide scale is the combinatorial explosion of SNP interactions. The focus on many current studies with MDR is on scaling this approach to genome-wide association data. Other than faster computer hardware for MDR (Greene *et al.*, 2010c; Sinnott-Armstrong *et al.*, 2009), or parallel implementations (Bush *et al.*, 2006), there are two general strategies that are being pursued, each of which has advantages and disadvantages. The first is a filter approach that preselects SNPs that are likely to interact prior to MDR analysis. The second approach is to use stochastic search algorithms, such as those reviewed by Michalewicz and Fogel (2004) to guide an MDR analysis. We review each of these approaches in turn.

III. FILTER APPROACHES TO MDR

As discussed above, it is computationally infeasible to combinatorially explore all interactions among the DNA sequence variations in a genome-wide association study. One approach is to filter out a subset of variations that can then be efficiently analyzed using a method such as MDR. We review below a powerful filter method based on the ReliefF algorithm and then discuss prospects for using biological knowledge to filter genetic variations.

A. ReliefF filters

There are many different statistical and computational methods for determining the quality of attributes. A standard strategy in human genetics and epidemiology is to assess the quality of each SNP using a chi-square test of independence followed by a correction of the significance level that takes into account an increased false-positive (i.e., type I error) rate due to multiple tests. This is a very efficient filtering method but it ignores the dependencies or interactions between genes. Kira and Rendell (1992) developed an algorithm called Relief that is capable of detecting attribute dependencies. Relief estimates the quality of attributes through a type of nearest neighbor algorithm that selects neighbors (instances) from the same class and from the different class based on the vector of values across attributes. Weights (W) or quality estimates for each attribute (A) are estimated based on whether the nearest neighbor (nearest hit, H) of a randomly selected instance (R) from the same class and the nearest neighbor from the other class (nearest miss, M) have the same or different values. This process of adjusting weights is repeated for m instances. The algorithm produces weights for each attribute ranging from -1 (worst) to $+1$ (best). The Relief pseudocode is outlined below:

```
set all weights W[A] = 0
for i = 1 to m do begin
        randomly select an instance R_i
        find nearest hit H and nearest miss M
        for A = 1 to a do
                W[A] = W[A] – diff(A, R_i, H)/m + diff(A, R_i, M)/m
        end
```

The function diff(A, I_1, I_2) calculates the difference between the values of the attribute A for two instances I_1 and I_2. For nominal attributes such as SNPs it is defined as:

$$\text{diff}(A, I_1, I_2) = 0 \text{ if genotype}(A, I_1) = \text{genotype}(A, I_2), \quad 1 \text{ otherwise}$$

The time complexity of Relief is $O(m^*n^*a)$ where m is the number of instances randomly sampled from a dataset with n total instances and a attributes. Kononenko (1994) improved upon Relief by choosing n nearest neighbors instead of just one. This new ReliefF algorithm has been shown to be more robust to noisy attributes (Kononenko, 1994; Robnik-Siknja and Kononenko, 2001, 2003) and is widely used in data mining applications.

Several modifications and extensions of the ReliefF algorithm have been proposed for genome-wide analysis. Moore and White (2007) proposed a "tuned" ReliefF algorithm (TuRF) that systematically removes attributes that have low quality estimates so that the ReliefF values if the remaining attributes can be re-estimated. McKinney *et al.* (2007) proposed a hybrid ReliefF algorithm that uses measures of entropy to boost performance. Greene *et al.* (2009) developed a spatially uniform ReliefF (SURF) algorithm that uses a fixed distance for picking neighbors. Greene *et al.* (2010b) proposed a SURF* algorithm that extends SURF by also using information from the furthest neighbors. All of these methods have demonstrated the ability to successfully filter interacting SNPs in an efficient manner. It is important to note that other methods such as random forests have also been used to preprocess SNPs for gene–gene interaction analysis (De Lobel *et al.*, 2010). The ReliefF algorithms described here are available in the open-source MDR software package available on http://sourceforge.net.

B. Biological filters

ReliefF and other measures such as interaction information (Moore *et al.*, 2006) are likely to be very useful for providing an analytical means for filtering genetic variations prior to epistasis analysis using decision trees or MDR, for example. However, there is growing recognition that we should use the wealth of accumulated knowledge about gene function to prioritize which genetic variations are analyzed for gene–gene interactions. For any given disease there are often multiple biochemical pathways, for example, that have been experimentally confirmed to play an important role in the etiology of the disease. Genes in these pathways can be selected for gene–gene interaction analysis thus significantly reducing the number of gene–gene interaction tests that need to be performed. Gene Ontology (GO), chromosomal location and protein–protein interactions are all example sources of expert knowledge that can be used in a similar manner. Pattin and Moore (2008) have specifically reviewed protein–protein interaction databases as a source of expert knowledge that can be used to guide genome-wide association studies of epistasis. Bush *et al.* (2009) have presented a Biofilter algorithm for integrating expert knowledge for gene–gene interaction analysis. Askland *et al.* (2009) have shown how the exploratory visual analysis (EVA) method can be used to select SNPs in specific pathways and GO groups. Additional work in this area is needed.

IV. WRAPPER APPROACHES TO MDR

Stochastic search or wrapper methods may be more powerful than filter approaches because no attributes are discarded in the process. As a result, every attribute retains some probability of being selected for evaluation by the classifier. There are many different stochastic wrapper algorithms that can be applied to this problem (Michalewicz and Fogel, 2004). However, when interactions are present in the absence of marginal effects, there is no reason to expect that any wrapper method would perform better than a random search because there are no "building blocks" for this problem when accuracy is used as the fitness measure. That is, the fitness of any given classifier would look no better than any other with just one of the correct SNPs in the MDR model. Preliminary studies by White *et al.* (2005) support this idea. For GP or any other wrapper to work there needs to be recognizable building blocks. Greene *et al.* (2008) have explored the use of estimation of distribution algorithms (EDAs). The general idea of EDAs is that solutions to a problem are statistically modeled and the ensuing probability distribution function is used to generate new models. In this initial study, the authors compared an EDA that utilizes only accuracy from an MDR classifier to update the probabilities with which SNPs are selected with an EDA that uses both accuracy and preprocessed TuRF scores. The results indicated that the EDA approach that used TuRF scores as expert knowledge significantly outperformed the approach that just used MDR accuracies. This is because the TuRF scores provide the building blocks that are needed to point the search algorithm in the right direction. More recent studies have focused on the optimal use of expert knowledge in the EDA algorithm (Greene *et al.*, 2009). The EDA algorithms described here are available in the open-source MDR software package available on http://sourceforge.net.

V. STATISTICAL INTERPRETATION OF MDR MODELS

The MDR method described above is a powerful attribute construction approach for detecting epistasis or nonlinear gene–gene interactions in epidemiologic studies of common human diseases. The models that MDR produces are by nature multidimensional and thus difficult to interpret. For example, an interaction model with four SNPs, each with three genotypes, summarizes 81 different genotype (i.e., level) combinations (i.e., 3^4). How do each of these level combinations relate back to biological processes in a cell? Why are some combinations associated with high risk for disease and some associated with low risk for disease? Moore *et al.* (2006) have proposed using information theoretic approaches with graph-based models to provide both a statistical and a visual interpretation of a multidimensional MDR model. Statistical interpretation should facilitate biological interpretation because it provides a deeper understanding of the relationship between the attributes and the class variable. We describe next the concept of interaction information and how it can be used to facilitate statistical interpretation.

Jakulin and Bratko (2003) have provided a metric for determining the gain in information about a class variable (e.g., case-control status) from merging two attributes into one (i.e., attribute construction) over that provided by the attributes independently. This measure of *information gain* allows us to gauge the benefit of considering two (or more) attributes as one unit. While the concept of information gain is not new (McGill, 1954), its application to the study of attribute interactions has been the focus of a recent study (Jakulin and Bratko, 2003). Consider two attributes, A and B, and a class label C. Let H(X) be the Shannon entropy (see Pierce, 1980) of X. The information gain (IG) of A, B, and C can be written as Eq. (5.1) and defined in terms of Shannon entropy (5.2 and 5.3).

$$\mathrm{IG(ABC) = I(A;B|C) - I(A;B)} \tag{5.1}$$

$$\mathrm{I(A;B|C) = H(A|C) + H(B|C) - H(A,B|C)} \tag{5.2}$$

$$\mathrm{I(A;B) = H(A) + H(B) - H(A,B)} \tag{5.3}$$

The first term in Eq. (5.1), I(A;B | C), measures the *interaction* of A and B. The second term, I(A;B), measures the *dependency* or correlation between A and B. If this difference is positive, then there is evidence for an attribute interaction that cannot be linearly decomposed. If the difference is negative, then the information between A and B is redundant. If the difference is zero, then there is evidence of conditional independence or a mixture of synergy and redundancy.

These measures of interaction information can be used to construct interaction graphs (i.e., network diagrams) and an interaction dendrograms using the entropy estimates from Step 1 with the algorithms described first by Jakulin and Bratko (2003) and more recently in the context of genetic analysis by Moore *et al.* (2006). Interaction graphs are comprised of a node for each attribute with pairwise connections between them. The percentage of entropy removed (i.e., information gain) by each attribute is visualized for each node. The percentage of entropy removed for each pairwise MDR product of attributes is visualized for each connection. Thus, the independent main effects of each polymorphism can be quickly compared to the interaction effect. Additive and nonadditive interactions can be quickly assessed and used to interpret the MDR model which consists of distributions of cases and controls for each genotype combination. Positive entropy values indicate synergistic interaction while negative entropy values indicate redundancy.

Interaction dendrograms are also a useful way to visualize interaction (Jakulin and Bratko, 2003; Moore *et al.*, 2006). Here, hierarchical clustering is used to build a dendrogram that places strongly interacting attributes close together at the leaves of the tree. Jakulin and Bratko (2003) define the following dissimilarity measure, D (Eq. 5.4), that is used by a hierarchical clustering algorithm to build a dendrogram. The value of 1000 is used as an upper bound to scale the dendrograms.

$$D(A,B) = |I(A;B;C)|^{-1} \quad \text{if } |I(A;B;C)|^{-1} < 1000 \\ 1000 \text{ otherwise} \tag{5.4}$$

Using this measure, a dissimilarity matrix can be estimated and used with hierarchical cluster analysis to build an interaction dendrogram. This facilitates rapid identification and interpretation of pairs of interactions. The algorithms for the entropy-based measures of information gain are implemented in the open-source MDR software package available from www.epistasis.org. Output in the form of interaction dendrograms and interaction graphs is provided.

VI. SUMMARY

As human genetics and epidemiology move into the genomics age with access to all the information in the genome, we will become increasingly dependent on computer science for managing and making sense of these mountains of data. The specific challenge reviewed here is the detection, characterization, and interpretation of epistasis or gene–gene interactions that are predictive of susceptibility to common human diseases. Epistasis is an important source of complexity in the genotype to phenotype map that requires special computational methods for analysis. We have reviewed a powerful attribute construction method called multifactor dimensionality reduction or MDR that can be used in a classification framework to detect nonlinear attribute interactions in genetic studies of common human diseases. We have also reviewed a filter method using ReliefF and a stochastic wrapper method using EDAs for the analysis of gene–gene interaction on a genome-wide scale with hundreds of thousands of attributes. Finally, we reviewed information theoretic methods to facilitate the statistical and subsequent biological interpretation of high-order gene–gene interaction models. These data mining and knowledge discovery methods and others will play an increasingly important role in human genetics as the field moves away from the candidate-gene approach that focuses on a few targeted genes to the genome-wide approach that measures DNA sequence variations from across the genome. We anticipate the role of expert biological knowledge will be increasingly important for successful data mining and thus learning algorithms need to adapt to exploit this information.

Acknowledgments

This work was supported by National Institutes of Health (USA) grants LM009012, LM010098, and AI59694.

References

Andrew, A. S., Nelson, H. H., Kelsey, K. T., Moore, J. H., Meng, A. C., Casella, D. P., Tosteson, T. D., Schned, A. R., and Karagas, M. R. (2006). Concordance of multiple analytical approaches demonstrates a complex relationship between DNA repair gene SNPs, smoking, and bladder cancer susceptibility. *Carcinogenesis* **27**, 1030–1037.

Andrew, A. S., Karagas, M. R., Nelson, H. H., Guarrera, S., Polidoro, S., Gamberini, S., Sacerdote, C., Moore, J. H., Kelsey, K. T., Demidenko, E., Vineis, P., and Matullo, G. (2008). DNA repair polymorphisms modify bladder cancer risk: A multi-factor analytic strategy. *Hum. Hered.* **65**, 105–118.

Askland, K., Read, C., and Moore, J. (2009). Pathways-based analyses of whole-genome association study data in bipolar disorder reveal genes mediating ion channel activity and synaptic neurotransmission. *Hum. Genet.* **125**, 63–79.

Bateson, W. (1909). Mendel's Principles of Heredity. Cambridge University Press, Cambridge.

Bush, W. S., Dudek, S. M., and Ritchie, M. D. (2006). Parallel multifactor dimensionality reduction: A tool for the large-scale analysis of gene–gene interactions. *Bioinformatics* **22**, 2173–2174.

Bush, W. S., Edwards, T. L., Dudek, S. M., McKinney, B. A., and Ritchie, M. D. (2008). Alternative contingency table measures improve the power and detection of multifactor dimensionality reduction. *BMC Bioinform.* **9**, 238.

Bush, W. S., Dudek, S. M., and Ritchie, M. D. (2009). Biofilter: A knowledge-integration system for the multi-locus analysis of genome-wide association studies. *Pac. Symp. Biocomput.* 368–379.

Calle, M. L., Urrea, V., Vellalta, G., Malats, N., and Steen, K. V. (2008). Improving strategies for detecting genetic patterns of disease susceptibility in association studies. *Stat. Med.* **27**, 6532–6546.

Calle, M. L., Urrea, V., Malats, N., and Van Steen, K. (2010). mbmdr: An R package for exploring gene–gene interactions associated with binary or quantitative traits. *Bioinformatics* (in press).

Cattaert, T., Urrea, V., Naj, A. C., De Lobel, L., De Wit, V., Fu, M., Mahachie John, J. M., Shen, H., Calle, M. L., Ritchie, M. D., Edwards, T. L., and Van Steen, K. (2010). FAM-MDR: A flexible family-based multifactor dimensionality reduction technique to detect epistasis using related individuals. *PLoS ONE* **5**, e10304.

Chung, Y., Lee, S. Y., Elston, R. C., and Park, T. (2007). Odds ratio based multifactor-dimensionality reduction method for detecting gene–gene interactions. *Bioinformatics* **23**, 71–76.

Cordell, H. J. (2009). Genome-wide association studies: Detecting gene–gene interactions that underlie human diseases. *Nat. Rev. Genet.* **10**, 392–404.

Culverhouse, R., Suarez, B. K., Lin, J., and Reich, T. (2002). A perspective on epistasis: Limits of models displaying no main effect. *Am. J. Hum. Genet.* **70**, 461–471.

De Lobel, L., Geurts, P., Baele, G., Castro-Giner, F., Kogevinas, M., and Van Steen, K. (2010). A screening methodology based on Random Forests to improve the detection of gene–gene interactions. *Eur. J. Hum. Genet.* (in press).

Edwards, T. L., Turner, S. D., Torstenson, E. S., Dudek, S. M., Martin, E. R., and Ritchie, M. D. (2010). A general framework for formal tests of interaction after exhaustive search methods with applications to MDR and MDR-PDT. *PLoS ONE* **5**, e9363.

Fisher, R. A. (1918). The correlations between relatives on the supposition of Mendelian inheritance. *Trans. R Soc. Edinb.* **52**, 399–433.

Greene, C. S., Penrod, N. M., Kiralis, J., and Moore, J. H. (2008). Spatially uniform ReliefF (SURF) for computationally-efficient filtering of gene–gene interactions. *BioData Min.* **2**, 5.

Greene, C. S., Gilmore, J., Kiralis, J., Andrews, P. C., and Moore, J. H. (2009). Optimal use of expert knowledge in ant colony optimization for the analysis of epistasis in human disease. *Lect. Notes Comput. Sci.* **5483**, 92–103.

Greene, C. S., Himmelstein, D. S., Nelson, H. H., Kelsey, K. T., Williams, S. M., Andrew, A. S., Karagas, M. R., and Moore, J. H. (2010a). Enabling personal genomics with an explicit test of epistasis. *Pac. Symp. Biocomput.* 327–336.

Greene, C. S., Himmelstein, D. S., Kiralis, J., and Moore, J. H. (2010b). The informative extremes: Using both nearest and farthest individuals can improve Relief algorithms in the domain of human genetics. *Lect. Notes Comput. Sci.* **6023,** 182–193.

Greene, C. S., Sinnott-Armstrong, N. A., Himmelstein, D. S., Park, P. J., Moore, J. H., and Harris, B. T. (2010c). Multifactor dimensionality reduction for graphics processing units enables genome-wide testing of epistasis in sporadic ALS. *Bioinformatics* **26,** 694–695.

Gui, J., Andrew, A. S., Andrews, P., Nelson, H. H., Kelsey, K. R., Karagas, M. R., and Moore, J. H. (2010a). A simple and computationally-efficient sampling approach to covariate adjustment for multifactor dimensionality reduction analysis of epistasis. *Hum. Hered.* (in press).

Gui, J., Andrew, A. S., Andrews, P., Nelson, H. H., Kelsey, K. R., Karagas, M. R., and Moore, J. H. (2010b). A robust multifactor dimensionality reduction method for detecting gene–gene interactions with application to the genetic analysis of bladder cancer susceptibility. *Ann. Hum. Genet.* (in press).

Hahn, L. W., Ritchie, M. D., and Moore, J. H. (2003). Multifactor dimensionality reduction software for detecting gene–gene and gene–environment interactions. *Bioinformatics* **19,** 376–382.

Jakulin, A., and Bratko, I. (2003). Analyzing attribute interactions. *Lect. Notes Comput. Sci.* **2838,** 229–240.

Kira, K., and Rendell, L. A. (1992). A practical approach to feature selection. Proceedings of the Ninth International Workshop on Machine Learning, pp. 249–256.

Kononenko, I. (1994). Estimating attributes: Analysis and extension of Relief. Proceedings of the European Conference on Machine Learning, pp. 171–182.

Lee, S. Y., Chung, Y., Elston, R. C., Kim, Y., and Park, T. (2007). Log-linear model-based multifactor dimensionality reduction method to detect gene–gene interactions. *Bioinformatics* **23,** 2589–2595.

Lewontin, R. C. (1974). The analysis of variance and the analysis of causes. *Am. J. Hum. Genet.* **26,** 400–411.

Li, W., and Reich, J. (2000). A complete enumeration and classification of two-locus disease models. *Hum. Hered.* **50,** 334–349.

Lou, X. Y., Chen, G. B., Yan, L., Ma, J. Z., Zhu, J., Elston, R. C., and Li, M. D. (2007). A generalized combinatorial approach for detecting gene-by-gene and gene-by-environment interactions with application to nicotine dependence. *Am. J. Hum. Genet.* **80**(6), 1125–1137.

Lou, X. Y., Chen, G. B., Yan, L., Ma, J. Z., Mangold, J. E., Zhu, J., Elston, R. C., and Li, M. D. (2008). A combinatorial approach to detecting gene–gene and gene–environment interactions in family studies. *Am. J. Hum. Genet.* **83,** 457–467.

Martin, E. R., Ritchie, M. D., Hahn, L., Kang, S., and Moore, J. H. (2006). A novel method to identify gene–gene effects in nuclear families: The MDR-PDT. *Genet. Epidemiol.* **30,** 111–123.

McGill, W. J. (1954). Multivariate information transmission. *Psychometrica* **19,** 97–116.

McKinney, B. A., Reif, D. M., Ritchie, M. D., and Moore, J. H. (2006). Machine learning for detecting gene–gene interactions: A review. *Appl. Bioinformatics* **5,** 77–88.

McKinney, B. A., Reif, D. M., White, B. C., Crowe, J. E., Jr., and Moore, J. H. (2007). Evaporative cooling feature selection for genotypic data involving interactions. *Bioinformatics* **23,** 2113–2120.

Mei, H., Cuccaro, M. L., and Martin, E. R. (2007). Multifactor dimensionality reduction-phenomics: A novel method to capture genetic heterogeneity with use of phenotypic variables. *Am. J. Hum. Genet.* **81,** 1251–1261.

Michalewicz, Z., and Fogel, D. B. (2004). How to Solve It: Modern Heuristics. Springer, New York.

Michalski, R. S. (1983). A theory and methodology of inductive learning. *Artif. Intell.* **20,** 111–161.

Millstein, J., Conti, D. V., Gilliland, F. D., and Gauderman, W. J. (2006). A testing framework for identifying susceptibility genes in the presence of epistasis. *Am. J. Hum. Genet.* **78,** 15–27.

Moore, J. H. (2003). The ubiquitous nature of epistasis in determining susceptibility to common human diseases. *Hum. Hered.* **56,** 73–82.

Moore, J. H. (2004). Computational analysis of gene–gene interactions in common human diseases using multifactor dimensionality reduction. *Expert Rev. Mol. Diagn.* **4,** 795–803.

Moore, J. H. (2007). Genome-wide analysis of epistasis using multifactor dimensionality reduction: Feature selection and construction in the domain of human genetics. *In* "Knowledge Discovery and Data Mining: Challenges and Realities with Real World Data" (X. Zhu, ed.), pp. 17–30. IGI, Hershey.

Moore, J. H., and Ritchie, M. D. (2004). The challenges of whole-genome approaches to common diseases. *J. Am. Med. Assoc.* **291,** 1642–1643.

Moore, J. H., and White, B. C. (2007). Tuning ReliefF for genome-wide genetic analysis. *Lect. Notes Comput. Sci.* **4447,** 166–175.

Moore, J. H., and Williams, S. W. (2002). New strategies for identifying gene–gene interactions in hypertension. *Ann. Med.* **34,** 88–95.

Moore, J. H., and Williams, S. W. (2005). Traversing the conceptual divide between biological and statistical epistasis: Systems biology and a more modern synthesis. *BioEssays* **27,** 637–646.

Moore, J. H., and Williams, S. M. (2009). Epistasis and its implications for personal genetics. *Am. J. Hum. Genet.* **85,** 309–320.

Moore, J. H., Gilbert, J. C., Tsai, C.-T., Chiang, F. T., Holden, W., Barney, N., and White, B. C. (2006). A flexible computational framework for detecting, characterizing, and interpreting statistical patterns of epistasis in genetic studies of human disease susceptibility. *J. Theor. Biol.* **241,** 252–261.

Namkung, J., Elston, R. C., Yang, J. M., and Park, T. (2009a). Identification of gene–gene interactions in the presence of missing data using the multifactor dimensionality reduction method. *Genet. Epidemiol.* **33,** 646–656.

Namkung, J., Kim, K., Yi, S., Chung, W., Kwon, M. S., and Park, T. (2009b). New evaluation measures for multifactor dimensionality reduction classifiers in gene–gene interaction analysis. *Bioinformatics* **25,** 338–345.

Pattin, K. A., and Moore, J. H. (2008). Exploiting the proteome to improve the genome-wide genetic analysis of epistasis in common human diseases. *Hum. Genet.* **124,** 19–29.

Pattin, K. A., White, B. C., Barney, N., Gui, J., Nelson, H. H., Kelsey, K. T., Andrew, A. S., Karagas, M. R., and Moore, J. H. (2009). A computationally efficient hypothesis testing method for epistasis analysis using multifactor dimensionality reduction. *Genet. Epidemiol.* **33,** 87–94.

Phillips, P. C. (1998). The language of gene interaction. *Genetics* **149,** 1167–1171.

Phillips, P. C. (2008). Epistasis—The essential role of gene interactions in the structure and evolution of genetic systems. *Nat. Rev. Genet.* **9,** 855–867.

Pierce, J. R. (1980). An Introduction to Information Theory: Symbols, Signals, and Noise. Dover, New York.

Ritchie, M. D., Hahn, L. W., Roodi, N., Bailey, L. R., Dupont, W. D., Parl, F. F., and Moore, J. H. (2001). Multifactor dimensionality reduction reveals high-order interactions among estrogen metabolism genes in sporadic breast cancer. *Am. J. Hum. Genet.* **69,** 138–147.

Ritchie, M. D., Hahn, L. W., and Moore, J. H. (2003). Power of multifactor dimensionality reduction for detecting gene–gene interactions in the presence of genotyping error, phenocopy, and genetic heterogeneity. *Genet. Epidemiol.* **24,** 150–157.

Robnik-Siknja, M., and Kononenko, I. (2003). Theoretical and empirical analysis of ReliefF and RReliefF. *Mach. Learn.* **53,** 23–69.

Robnik-Sikonja, M., and Kononenko, I. (2001). Comprehensible interpretation of Relief's estimates. Proceedings of the Eighteenth International Conference on Machine Learning, pp. 433–440.

Sinnott-Armstrong, N. A., Greene, C. S., Cancare, F., and Moore, J. H. (2009). Accelerating epistasis analysis in human genetics with consumer graphics hardware. *BMC Res. Notes* **2,** 149.

Tyler, A. L., Asselbergs, F. W., Williams, S. M., and Moore, J. H. (2009). Shadows of complexity: What biological networks reveal about epistasis and pleiotropy. *Bioessays* **31,** 220–227.

Velez, D. R., White, B. C., Motsinger, A. A., Bush, W. S., Ritchie, M. D., Williams, S. M., and Moore, J. H. (2007). A balanced accuracy function for epistasis modeling in imbalanced datasets using multifactor dimensionality reduction. *Genet. Epidemiol.* **31,** 306–315.

Waddington, C. H. (1942). Canalization of development and the inheritance of acquired characters. *Nature* **150,** 563–565.

White, B. C., Gilbert, J. C., Reif, D. M., and Moore, J. H. (2005). A statistical comparison of grammatical evolution strategies in the domain of human genetics. Proceedings of the IEEE Congress on Evolutionary Computing, pp. 676–682.

6

The Restricted Partition Method

Robert Culverhouse
Washington University in St Louis School of Medicine, St Louis, Missouri, USA

Advances in Genetics, Vol. 72

0065-2660/10 $35.00
DOI: 10.1016/S0065-2660(10)72006-X

ABSTRACT

For many complex traits, the bulk of the phenotypic variation attributable to genetic factors remains unexplained, even after well-powered genome-wide association studies. Among the multiple possible explanations for the "missing" variance, joint effects of multiple genetic variants are a particularly appealing target for investigation: they are well documented in biology and can often be evaluated using existing data. The first two sections of this chapter discusses these and other concerns that led to the development of the Restricted Partition Method (RPM).

The RPM is an exploratory tool designed to investigate, in a model agnostic manner, joint effects of genetic and environmental factors contributing to quantitative or dichotomous phenotypes. The method partitions multilocus genotypes (or genotype-environmental exposure classes) into statistically distinct "risk" groups, then evaluates the resulting model for phenotypic variance explained. It is sensitive to factors whose effects are apparent only in a joint analysis, and which would therefore be missed by many other methods. The third section of the chapter provides details of the RPM algorithm and walks the reader through an example.

The final sections of the chapter discuss practical issues related to the use of the method. Because exhaustive pairwise or higher order analyses of many SNPs are computationally burdensome, much of the discussion focuses on computational issues. The RPM proved to be practical for a large candidate gene analysis, consisting of over 40,000 SNPs, using a desktop computer. Because the algorithm and software lend themselves to distributed processing, larger analyses can easily be split among multiple computers.

The Restricted Partition Method (RPM) is an exploratory analytic tool focused on identifying genetic or environmental factors contributing to phenotypic variance, including contributions that are primarily expressed through interactions and thus might otherwise be undetectable.

I. WHY LOOK FOR JOINT GENETIC EFFECTS?

Technological improvements in genotyping have made genome-wide association studies (GWAS) a practical reality. Traditional genetic analysis methods exploit the possibility that variation in a single genetic locus may result in detectable effects on a phenotype, and that at least one of the variants measured in the GWAS will have sufficient correlation to the disease locus to identify it. These methods typically examine genotypes one at a time, and build additive (or log-additive) effects models from them. The approach has been successful, with GWAS alone identifying over 500 genetic variants contributing to disease (Manolio *et al.*, 2009).

However, after hundreds of studies on scores of phenotypes, a discouraging pattern has emerged. Most common gene variants identified in these studies account for only a small fraction of the variation in phenotype predicted to be due to genetics, much less the overall phenotypic variance. (Hirschhorn, 2009; Kraft and Hunter, 2009; Manolio *et al.*, 2009)

As a result, a key challenge for genetic analysis today is to account for the bulk of the phenotypic variance in complex traits attributable to genetic factors. A variety of factors have been suggested to account for the "missing variance," including rare variants of large effect that have not been surveyed by current GWAS strategies, inflated heritability estimates, structural variants (e.g., copy number variants such as insertion/deletions or copy neutral variation such as inversions and translocations) (Manolio *et al.*, 2009), population heterogeneity, parent-of-origin effects, gene–gene interactions, and gene–environment interactions (Galvan *et al.*, 2010).

While all of these likely play some role in the "missing" genetic heritability for complex diseases, the possibility of joint effects (gene–gene or gene–environment) presents a particularly appealing target for research, in spite of the associated computational and analytical challenges.

First, we know that such mechanisms play an essential role in the structure and evolution of genetic systems (Phillips, 2008) and appear to be correlated with genomic complexity (Sanjuan and Elena, 2006). Biology is replete with well-documented examples of epistasis and gene–environment interactions affecting phenotypes ranging from gross morphology to longevity to efficiency of reproduction (Anholt *et al.*, 2003; Gerke *et al.*, 2009; Mackay, 2010; Vieira *et al.*, 2000; Wolf *et al.*, 2005). Further, most traits of medical interest (such as heart disease, hypertension, diabetes, cancer, and infection) arise from biological systems controlled by interacting genetic factors (Churchill *et al.*, 2004; Lander and Schork, 1994; Phillips, 2008; Routman and Cheverud, 1995; Schork, 1997; Sing *et al.*, 2003; Szathmary *et al.*, 2001).

The second appealing feature of using joint-effect analyses to try to account for additional genetic variance is that it is the one explanatory factor that can be readily examined in existing data, which has typically been collected at great expense. The other possible explanations typically require additional genotyping or sequencing (rare variants, structural variants), other biochemical tests (epigenetic effects), or gathering new samples (re-estimating heritability, parent of origin effects), possibly resulting in considerable delay and increased cost.

II. KEY CONCERNS ADDRESSED BY THE RPM

The RPM addresses several concerns that arise when attempting to account for additional phenotypic variance by examining factors jointly.

A. The possibility of joint effects with small marginal effects

Researchers who study model organisms can manipulate allele frequencies to interrogate the effects (including interactions) of particular variants. In natural populations, such as humans, it can be difficult to develop a model for a single variant that generalizes well across samples, much less to predict how interactions will be expressed. One possibility that could account for some of the "missing variance" is that interactions of substantial effect may involve genetic loci that do not display noticeable marginal effects.

Interactions of this kind have been reported for many years. In one example related to triglyceride levels, two diallelic loci (an InDel in *APOAl-C3-A4* and a RFLP in *LDLR*) jointly accounted for 9% of the trait variability, even though individually the InDel only accounted for 1% of the variability and the RFLP accounted for less than 0.1% (Nelson *et al.*, 2001).

Subsequently, theoretical work on epistatic models quantified exactly how much phenotypic variation could be due to a small number of interacting factors (e.g., SNPs) which would be very difficult to detect if examined one at a time (Culverhouse *et al.*, 2002). This work showed that even for moderate to low frequency phenotypes, a handful of SNPs with negligible marginal effects could account for nearly all the phenotypic variance. This suggested that joint analyses should not be limited to evaluating possible interaction among factors demonstrating significant marginal effects, even though such a strategy is sometimes advocated. On the other hand, the theoretical work also showed that it was not possible to "hide" a large contribution to phenotype in interactions that could *only* be seen with a high order joint analysis.

Figure 6.1 illustrates this idea using a dichotomous trait, three loci, and allele frequencies = 0.5. The horizontal axis represents the phenotype prevalence and the vertical axis represents the maximum proportion of phenotypic variance attributable to genotype in a model where no individual locus has a marginal association to the phenotype (i.e., a purely epistatic model). The top curve represents the maximum variance that can be explained by genotype in a three-locus purely epistatic model. The bottom curve represents the maximum variance that can be explained in a three-locus purely epistatic model where the loci could not be detected with two-locus analyses. For reference, the intermediate curve represents the maximum variance in phenotype that can be explained using a two-locus purely epistatic model.

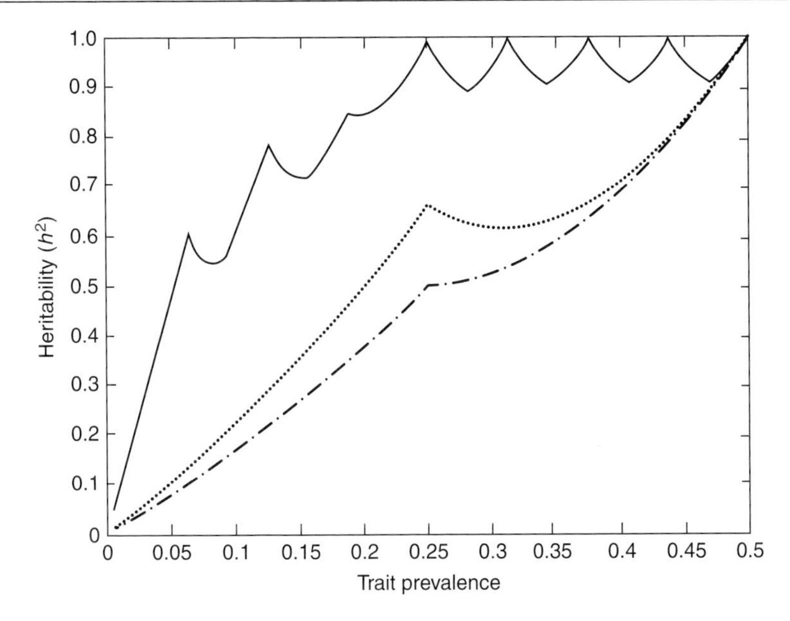

Figure 6.1. Comparison of maximum heritabilities for three-locus, purely epistatic models with (top curve) and without (bottom curve) two-locus interactions. The maximum heritabilities for two-locus, purely epistatic models (middle curve) are included as a reference.

These theoretical results suggest that there may be an analytic "sweet spot" for detecting factors that contribute to phenotype variation, but would likely be missed if SNPs were only analyzed individually: examining variants as pairs (or triples) might be able to identify factors that were part of complex systems, but which could potentially be very difficult to identify with univariate analyses.

B. Model agnosticism

There is a trade-off that researchers make whenever they choose a particular modeling approach for an analysis. If they pick a highly parameterized model with many distributional assumptions, and the true biological system fits that model well, they can achieve unsurpassed statistical power. Unfortunately, when the assumptions are violated, depending on the distribution of the data, such models can lead to results that appear either much more significant than they really are (false positives) or results that appear less significant than they are in reality (false negatives). In practice, researchers must balance these two types of errors. As recent rapid technological improvements have opened the floodgates on the production of genome-wide genetic data (tens of thousands, then hundreds of thousands, then over a million variants genotyped for each subject in a genome-wide scan), analysts and traditional analytic methods have been overwhelmed.

The first, reasonable response was to perform fast analyses: assume all variants act additively and independently and retain signals whose *p*-values pass a threshold related to the estimated number of independent tests possible based on common variants (minor allele frequency (MAF) $\geq 5\%$) in the human genome. Various thresholds have been suggested, with 5×10^{-8} being commonly used (Risch and Merikangas, 1996). The idea was that since many of the genetic effects we had previously detected either appeared to be additive, or could have been detected with an additive test, and since we were surveying so many variants, this strategy might be expected to miss a few factors of low explanatory power but to identify variants accounting for a large portion of the phenotypic variance for common diseases. Unfortunately, so far that has not proven to be the case. While numerous variants associated to common complex diseases have been identified using this approach and well-replicated in multiple, large samples, such variants (as noted above) only account for a small portion of the phenotypic variance for common diseases.

In the effort to account for more genetic variance, before moving to multilocus models one might consider alternatives to the additive model for single locus analysis. If the true genetic model is dominant or recessive, an additive test still has a good chance of detecting the locus. Moreover, many researchers make it a habit to triple their tests by running all three of these models of action. This has the drawback of multiplying the number of tests performed by three, and still not providing an adequate fit if the true model involves overdominance, where the heterozygote phenotype does not lie between that of the two homozygotes. (The classic example of overdominance in humans is the fitness of the major sickle cell allele, HbS, in a malarial environment.) An alternative that will be sensitive to any of these models, with some loss of power compared to correctly guessing the true model *a priori*, is to use a full two degree of freedom model in the analysis.

Following the same line of thought, since the ultimate goal is to account for phenotypic variance and since it is unknown how the structure of high-order interactions will appear if only a few of the factors are modeled, it seems reasonable to be agnostic to a specific model of action for predictive factors (whether examined individually or jointly) in our analyses.

The focus on model agnosticism with "sensitivity to situations where interactions are present" has led some to think that the RPM as a test for interactions. This is not the case. The goal is to identify factors that may contribute to phenotypic variation, whether the contribution is expressed in a typical mode (e.g., an independent additive effect) or a more unusual mode (e.g., over dominance or an interaction). The RPM does not currently include a significance test for interactions, although such a test could be added. The method suggested by Edwards *et al.* (2010) is one approach that could practically be incorporated into the RPM software.

Although formal tests of interactions are important in many situations, undue emphasis on them can be counter-productive in grappling with complex human disease. Two examples can help illustrate this point.

1. Two examples

The first is a theoretical illustration of the fact that the magnitude of an interaction effect in a genetic analysis is not an inherent property of the genetic model, but depends on allele frequencies in the data. Templeton (2000) constructed a two-locus genetic model for total serum cholesterol (TSC) consistent with previously reported data indicating an interaction between *ApoE* and *LDLR* (Pedersen and Berg, 1989; Pedersen and Berg, 1990). Analysis of this model based on European allele frequencies (*ApoE* frequencies of 0.078, 0.77, and 0.152 for ε2, ε3, and ε4, respectively, and *LDLR* frequencies of 0.22 and 0.78 for A_1 and A_2, respectively), would suggest that *ApoE* is the "major gene" for TSC, accounting for 77.7% of the genetic variance (52.8 mg^2/dl^2), while *LDLR* is a minor player, accounting for only 5.5% of the genetic variance. The remaining 16.8% of the genetic variance could be attributed to an interaction, and would only be detected if the variants were analyzed jointly.

Templeton then evaluated the same genetic model in a hypothetical population with different allele frequencies (*ApoE* frequencies of 0.02, 0.03, and 0.95 for ε2, ε3, and ε4, respectively, and *LDLR* frequencies of 0.5 for both A_1 and A_2). The analysis of this hypothetical population indicated that that *ApoE* was only a minor contributor to the trait (accounting for 11.9% of the genetic variance); that *LDLR* was the "major gene" locus (accounting for 81.4% of the variance); and that the interaction is not very strong (accounting for only 6.4% of the variance).

These results suggest that when studying complex diseases, it may be more important to focus on identifying contributors to trait variation than to focus on estimating particular parameters, such as the size of the interaction term.

A second example of how one could miss something important by focusing too much on the significance of an interaction parameter was found in a recent examination of data on smoking. Standard univariate analysis identified an association between nicotine dependence and a nonsynonymous coding SNP, rs16969968, in *CHRNA5*, a nicotinic receptor sub-unit gene. The RPM also identified this SNP as highly significant (accounting for 1.22% of the phenotypic variation). However, in a pairwise analysis of nicotinic receptor SNPs, the RPM identified a second nearby variant, rs3743075, which was not significant on its own (either by the RPM or a standard univariate analysis), but for which when combined with rs16969968 accounted for 1.83% of the variance. A logistic regression using this pair of predictors, however, indicated that rs3743075 was not significant on its own ($p=0.36$), nor was the interaction ($p=0.27$).

However, even without an interaction term, the logistic model including both SNPs accounts for 25% more of the trait variation than the sum of the two univariate effects. In fact, each SNP became more significant in the joint analysis than it was in a univariate analysis.

The reason for this surprising result is that the risk alleles for rs16969968 and rs3743075 are negatively correlated. Since rs16969968 has a bigger effect than rs3743075, when rs3743075 is analyzed on its own, its effect is almost completely masked by rs16969968. When rs16969968 is analyzed alone, its effect is dampened a bit by rs3743075, but still shows through.

In this case, requiring either a significant main effect or a significant interaction effect would have resulted in a failure to identify the contribution of rs3743075 to nicotine dependence. As it turns out, though rs3743075 is not associated with nicotine dependence in univariate analyses, it is associated with expression for *CHRNA5*, the gene for which rs16969968 alters the protein.

C. Computational complexity

The Combinatorial Partition Method (CPM; Nelson *et al.*, 2001) is a model agnostic method to examine genetic loci for their joint contribution to phenotypic variance which is sensitive even if the marginal effects from one or more of the loci are negligible. Its goal is to find a way to divide multilocus genotypes into subgroups in such a way that the grouping explains a large portion of the overall trait variation. It inspired the RPM as well as other methods such as the Multifactor Dimensionality Reduction (MDR) method (Hahn *et al.*, 2003).

To illustrate, consider a pair of biallelic loci corresponding to nine two-locus genotypes (AABB, AABb, AAbb, AaBB, AaBb, Aabb, aaBB, aaBb, aabb). The CPM analyzes the data in an intuitively straightforward but computationally intensive manner. Each possible partition of multilocus genotypes is evaluated for the amount of variation explained. The number of ways to partition g cells into k groups is known as a Sterling number of the second kind (Comtet, 1974) and is given by the formula:

$$S(g,k) = \frac{1}{k!}\sum_{i=0}^{k-1}(-1)^i\binom{k}{i}(k-i)^g$$

For example, this formula tells us that there are 255 ways to partition the nine genotypes into two groups, 3025 ways to partition nine genotypes into three groups, etc. Summing these, one finds that for any pair of biallelic loci, there are 21,146 ways to partition the genotypes into 2–9 groups. When there are n candidate loci involved, these 21,146 evaluations will need to be performed for each of the $n(n-1)/2$ ways to select two loci from the candidates. For 10 candidates, there are 45 pairs of loci to be evaluated, resulting in 951,570 two-locus model evaluations. Although this involves a great number of tests for only 10 loci, the fact that it could feasibly be done was demonstrated in the chapter that introduced the CPM (Nelson *et al.*, 2001). Using serum triglyceride levels as a phenotype, the authors detected clinically interesting interactions between loci that individually showed little or no effect on the phenotype. Although 2-way interactions can be analyzed with this method, its computational burden becomes overwhelming in the modern era of genotyping, where hundreds of thousands to millions of SNPs are routinely genotyped for a GWAS.

An additional computational concern is that, even if the CPM were restricted to a small number of candidate SNPs, the method does not scale well to models with more than two SNPs: the number of partitions possible with three biallelic loci is over 10^{21}. Clearly, if interactions involving more than two loci were to be analyzed, an alternative method needed to be developed.

The key observation leading to the RPM was that many of the partitions examined by the CPM did not need to be evaluated. Consider the distribution of a simulated trait presented in Fig. 6.2A. For five of the nine genotypes, the quantitative trait is distributed normally with mean 2, while the remaining four genotypes have a mean $= -2$. The goal of the CPM algorithm is to separate the genotypes into natural groups (in this case two groups). Figure 6.2B provides a good example of a partition that need not be evaluated. Genotype AABB has the lowest mean value for the quantitative trait, genotype aabb has the highest mean, and the remaining genotypes are intermediate between these two extremes. The partition that puts genotypes AABB and aabb into one group and the remaining genotypes into a second group does not need to be evaluated since it is guaranteed not to explain the variation optimally.

In general, to find the optimal partition, one need only rank the multilocus genotype groups by their trait value means and evaluate just those partitions which pool adjacent ranks. This immediately reduces the number of partitions that must be evaluated for a two-SNP model from 21,146 to

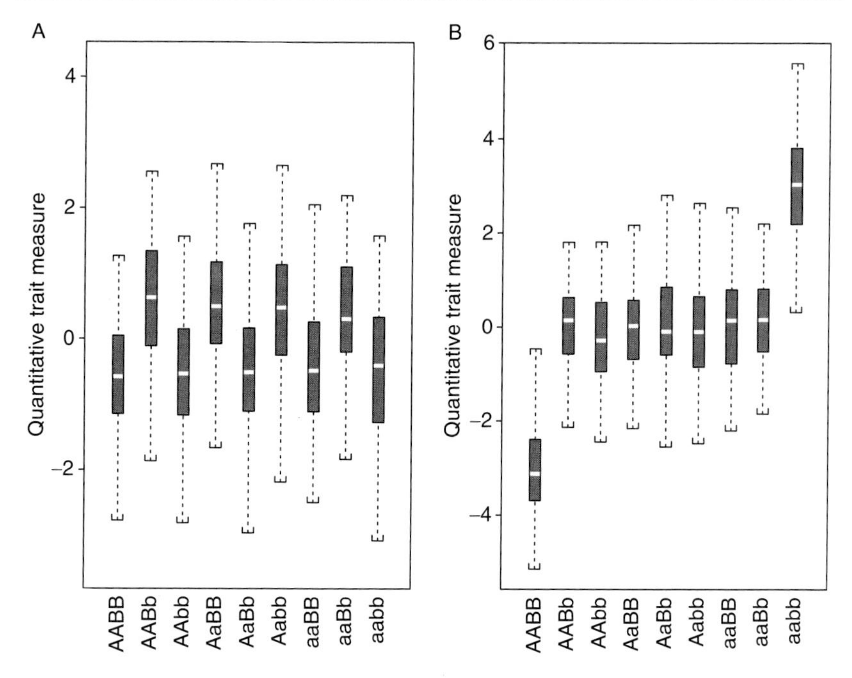

Figure 6.2. Two examples of phenotypic distributions varying with two-locus genotypes. (A) represents a purely epistatic model, where each genotype is associated with a high trait value or a low trait value. (B) illustrates the fact that partitions that group high and low trait value genotypes together will not be optimal to explain trait variance, and hence do not need to be evaluated.

$255 = 2^8 - 1$. For a 3-snp analysis, this approach reduces the partitions that must be evaluated from 10^{21} to $6.7 \times 10^7 = 2^{26} - 1$ (still not a very practical approach, but a considerable improvement).

D. Ameliorate the problem of overfitting

The merging of multiple genotypes into a single unit will typically reduce the explained variance (compared to treating each genotype as a distinct group). Why do this, if the goal is to account for phenotypic variance? The reason is that we hope to model biologically significant differences, not differences that may simply be due to sampling variability. The RPM aims to reduce the space of partitions to a single element for evaluation; and to do this in a way that provides some resistance to fitting the model to slight differences that arise merely from sampling variation. To do this, the RPM uses an agglomerative clustering strategy. At the beginning, each observed multilocus genotype is a terminal node (or leaf) for the tree. Two genotypes merge into the same subset of the partition if the phenotypic means for the two genotypes are not significantly different (the earliest join happens between the two nodes with closest means). By balancing maximization of the between group variation with minimization of the number of groups and the within group variation, the final model should have some protection from attempting to explain phenotypic differences that were likely due to sampling variability.

III. DESCRIPTION OF THE RPM

The RPM algorithm, described below, is an iterative search procedure for finding an optimized partition of the genotypes. Genotypes are sequentially merged based on the similarity of the mean values of their phenotypic trait. Selection of which genotypes to merge at each step is based on statistical criteria from a multiple comparisons test. Initially, each multilocus genotype forms its own group. The algorithm proceeds as follows:

1. A multiple comparisons test is performed to identify which (if any) groups have different mean quantitative trait values. The procedure halts if all groups have different means.
2. Pairs of genotype groups with means that are *not* significantly different from each other are ranked according to the difference in means between the two groups.
3. The pair from step 2 with the smallest difference (i.e., most similar mean values) is merged to form a new group.
4. The algorithm returns to step 1.

To provide a measure of the importance of the final results, the variance attributable to the joint genotypes in the final model (i.e., $R^2 =$ between group variance/total phenotypic variance) is computed. A natural consequence of this definition is that, if the genotypes are merged into a single group at the end of the algorithm, then $R^2 = 0$, reflecting the lack of evidence for quantitative trait differences between the genotypes. This does not indicate that the mean values of the genotypes are identical, only that the RPM does not have sufficient evidence to conclude that the differences are meaningful. How statistical significance is determined for a result when $R^2 > 0$ will be described later.

The selection of the multiple comparison test is arbitrary. We chose to use the Games–Howell variant of Tukey's Honestly Significant Difference (HSD) multiple comparison method with $\alpha = 0.05$ for our initial evaluation. (The Games–Howell version allows for the variance and sample sizes to vary between groups (Games and Howell, 1976).) Other methods we tested appear to perform similarly.

As a measure of the complexity of the algorithm, we note that each time the algorithm reaches Step 2, the number of groups is reduced by one. Thus, if there are initially n genotypes, the algorithm will always halt after no more than $n-1$ iterations.

A. An example of the algorithm in action

To illustrate, consider the case where the genotypes are derived from two biallelic loci. The individual data points will initially fall into one of nine groups (the two-locus genotypes), each with its own genotype-specific mean trait value and standard deviation. One such model is illustrated in Fig. 6.3. In each cell, the top number represents the mean trait value for the two-locus genotype. The bottom number (in parentheses) represents the standard deviation for the cell. In this figure, the marginal values (at the side and bottom of the table) represent the expected values that would be observed in a population where each of the loci had MAFs = 0.5.

Figure 6.4 is based on a sample of 1000 individuals drawn randomly from this population. As in Fig. 6.3, the top two numbers in each genotype cell represent the trait mean and standard deviation. The third number represents the number of subjects in each genotype. The unshaded cells in Fig. 6.4A indicate that none of the genotypes have yet been assigned to any subset of the partition.

The multiple comparison test divides the pairs into two groups: pairs with significantly different means and pairs with means that are not significantly different. If all nine two-locus genotypes (i.e., all 45 pairs) have significantly different means, the procedure halts and R^2 is computed for the nine-factor model (each cell is its own group). Otherwise, the algorithm proceeds to Step 2 where, of all the pairs whose means are not significantly different, the two with the smallest difference in means will be merged. The newly merged group will have a new mean and standard deviation. In Fig. 6.4A, the two genotypes with the closest means are AABB and Aabb. Since these means

	BB	Bb	bb	Marginal
AA	**0.0** (100.0)	**0.0** (100.0)	**75.0** (100.0)	**18.75** (105.1)
Aa	**0.0** (100.0)	**37.5** (100.0)	**0.0** (100.0)	**18.75** (101.7)
aa	**75.0** (100.0)	**0.0** (100.0)	**0.0** (100.0)	**18.75** (105.1)
Marginal	**18.75** (105.1)	**18.75** (101.7)	**18.75** (105.1)	**18.75** (103.5)

Figure 6.3. Generating Model: Mean trait values (standard deviation). All allele frequencies = 0.5. The genotypes are associated with three distinct levels for the trait.

were not found to be significantly different by our multiple comparisons test, they are merged as seen in Fig. 6.4B. As before, the eight remaining groups are assessed, and the groups with closest means that are not significantly different are merged. The process continues iteratively until the means of the remaining groups are significantly different from each other (Fig. 6.4C–G).

Since there were initially nine genotype groups, the final partition will be determined after no more than eight iterations of the algorithm (with a maximum of 120 pairwise comparisons of means) followed by one R^2 computation. In contrast, the CPM analysis of the same pair of loci would require R^2 to be calculated 21,146 times.

A

	BB	Bb	bb	
AA	**10.7** (103.6) *54*	**–6.8** (100.7) *128*	**98.6** (101.8) *52*	**20.6** (110.0) *234*
Aa	**12.9** (105.9) *116*	**40.7** (99.9) *265*	**10.5** (105.7) *125*	**26.9** (103.8) *506*
aa	**85.9** (107.3) *62*	**–1.5** (95.9) *136*	**–21.6** (86.7) *62*	**14.5** (104.9) *260*
	31.9 (110.7) *232*	**18.4** (101.6) *529*	**21.3** (109.0) *239*	

B

	BB	Bb	bb
AA	**10.6** (105.1)	**–6.8** (100.7) *128*	**98.6** (101.8) *52*
Aa	**12.9** (105.9) *116*	**40.7** (99.9) *265*	**10.6** (105.1)
aa	**85.9** (107.3) *62*	**–1.5** (95.9) *136*	**–21.6** (86.7) *62*

C

	BB	Bb	bb
AA	**11.5** (105.4)	**–6.8** (100.7) *128*	**98.6** (101.8) *52*
Aa	**11.5** (105.4)	**40.7** (99.9) *265*	**11.5** (105.4)
aa	**85.9** (107.3) *62*	**–1.5** (95.9) *136*	**–21.6** (86.7) *62*

Figure 6.4. (Continued)

D	BB	Bb	bb
AA	**11.5** (105.4)	**–4.1** (100.3)	**98.6** (101.8) *52*
Aa	**11.5** (105.4)	**40.7** (99.9) *265*	**11.5** (105.4)
aa	**85.9** (107.3) *62*	**–4.1** (100.3)	**–21.6** (86.7) *62*

E	BB	Bb	bb
AA	**4.2** (102.6)	**4.2** (102.6)	**98.6** (101.8) *52*
Aa	**4.2** (102.6)	**40.7** (99.9) *265*	**4.2** (102.6)
aa	**85.9** (107.3) *62*	**4.2** (102.6)	**–21.6** (86.7) *62*

F	BB	Bb	bb
AA	**4.2** (102.6)	**4.2** (102.6)	**91.7** (100.3)
Aa	**4.2** (102.6)	**40.7** (99.9) *265*	**4.2** (102.6)
aa	**91.7** (100.3)	**4.2** (102.6)	**–21.6** (86.7) *62*

G	BB	Bb	bb
AA	**1.6** (101.4)	**1.6** (101.4)	**91.7** (100.3)
Aa	**1.6** (101.4)	**40.7** (99.9)	**1.6** (101.4)
aa	**91.7** (100.3)	**1.6** (101.4)	**1.6** (101.4)

Figure 6.4. Example based on data simulated from the model in Fig. 6.3. $N = 1000$. Values in the cells are mean (standard deviation), and # *of subjects* in each group. (A) The algorithm begins by computing the mean and (standard deviation) of the trait value for each multilocus genotype. (B) The mean trait values for AABB and Aabb were not significantly different and they were the two closest means in the data, so the two genotypes

The computational difference is even more striking if three biallelic loci are analyzed jointly. The RPM would require at most 26 iterations (with no more than 3276 comparisons of means) to choose the partition of interest compared to more than 10^{21} partitions that would need to be evaluated using the CPM. For the RPM, given a sufficient sample size that the cells means could be estimated reasonably, analyzing possible 4-way interactions (maximum of 80 iterations) or more would be computationally feasible.

B. Permutation-based tests of significance

Because the R^2 estimated by this procedure has a distribution that is difficult to parameterize, the RPM software uses a permutation-based strategy to estimate *p*-values.

The basic idea of permutation-based significance tests is that an empirical null distribution can be generated from the data by breaking down any associations between the predictors and the test statistic by repeatedly permuting a key value or key values in the data. For instance, in our case we are interested in associations between genotypes and phenotypes. By permuting phenotype values among the individuals, the phenotype and genotype distributions are held constant (no matter whether genotypes are analyzed one at a time, two at a time, etc.). What we have done is to randomize the connection between the genotype and phenotype distributions. If we were to exhaustively go through every possible permutation of the data, we would have something similar to Fisher's exact test. Because there are typically so many possible permutations, we typically sample from the permutations using a random process.

For each permutation of the data, the test statistic is computed (in our case, the RPM model R^2). Because there should be no biological cause for a correlation between the genotype and phenotype in this permuted dataset, the

are merged. The combined group's mean and standard deviation are computed. (C) Genotype AaBB and the merged genotypes from (B) now have the minimal mean difference. The mean difference is not statistically significant, so they are merged. (D) Of the remaining groups, genotypes AABb and aaBb have the least mean difference. This difference is not significant, so the genotypes are merged. (E) At this point, the groups with the closest means are the one which includes the AABB genotype and the one that includes the AABb genotype. The means are not significantly different, so the groups are merged. (F) Of the remaining groups, genotypes AAbb and aaBB have the least mean difference. This difference is not significant, so the genotypes are merged. (G) Genotype aabb is merged with the group of genotypes including AABB. The differences in means between the three remaining groups are significant, so the procedure halts and the explained variance computed ($R^2 = 0.082$).

statistic is treated like a random roll of dice and it becomes one point in our empirically generated null. After repeating this process for many permutations, we get an idea of how this statistic would be distributed for data that had the given phenotypic and genotypic distributions, but the only correlation between the two was due to chance. By comparing our test statistic from the real, unpermuted data to this empirical null, we can evaluate how likely a result greater than or equal to ours would have arisen by chance.

Suppose our $R^2 = 0.012$. If, after 1000 permutations, we found that the R^2 from permuted data was ≥ 0.012 for 483 of the 1000, we would say that the test statistic had an empirical *p*-value of 0.48. (We generally like to report one decimal place less accuracy than the number of permutations performed to take sampling variability into account).

This general idea of using permutations to evaluate significance can be implemented in multiple ways. When the RPM software was written, the authors were concerned that a SNP with a MAF of 0.2 would likely have a different empirical null distribution than one with a MAF of 0.4; and it seemed even more likely that a pair of SNPs with MAF of 0.2 and 0.1 when analyzed jointly would have a different empirical null than two SNPs with MAF 0.4 and 0.5 (differences that would likely be exacerbated in an analysis of real data because missing data would also lead to differences in the phenotype distributions). For these reasons, the RPM software creates an individual empirical null for each analyzed unit (e.g., if there are SNPs s1, s2, s3, s4 are to be analyzed pairwise, there would be one empirical null generated for {s1,s2} one for {s1,s3}, one for {s1,s4}, . . . , and one for {s3,s4}. Simulation tests (Culverhouse *et al.*, 2004) indicate that if the SNPs in an analysis are not correlated, then a Bonferroni correction maintains the nominal experiment-wide error rate. Some of the computational consequences of our way of doing permutation testing will be discussed later in this chapter.

C. Appropriate for both quantitative and dichotomous traits

Although the design of the RPM focused on analyzing quantitative traits, the mechanics of the algorithm apply equally well to dichotomous traits coded by 0 and 1. To verify that this application of the method would provide meaningful results, extensive simulation testing was performed.

Initial analyses were performed on fully simulated data (both genotypes and phenotypes were simulated, genotypes in linkage equilibrium). Results from these tests of data generated from two-locus purely epistatic models indicated that the power to detect purely epistatic loci was good and the experimental rate of false positives in null data closely matched the nominal error rate. Further, a limited number of three-locus purely epistatic models were evaluated using two-locus analyses to test our prediction that the contributing loci would likely be identified. As expected, although none of the three loci could be detected using univariate analyses, all three loci were detected with good power using two-locus analyses, with no inflation in false positives.

In addition, the method was evaluated on a semi-simulated dataset (phenotypes simulated from a genetic model applied to real genotype data) provided by the Pharmacogenetics Research Network (PGRN) Analysis Workshop 2005. In these data, the RPM successfully identified a two-factor gene-by-environment interaction contributing to a dichotomous trait and the correct univariate contributor to quantitative trait from a block of SNPs in linkage disequilibrium (LD). (Culverhouse, 2007)

D. Designed for unrelated subjects, the RPM is potentially useful for small families

The RPM assumes that the subjects are unrelated. However, some researchers have suggested that including related individuals in an analysis intended for unrelated, as long as the family sizes are reasonably uniform, will only bias effect size estimates and not tend to create false positive signals. This suggestion was investigated as part of the Genetics Analysis Workshop (GAW) 16. This study involved a dataset of 42,461 real SNPs genotyped on 6479 family based subjects from the Framingham Heart Study (FHS) matched to multiple simulated phenotypes (Cupples *et al.*, 2009; Kraja *et al.*, 2009). From these, a subset of 1130 unrelated individuals was extracted for comparison. Since the phenotype model was simulated over real genotype data, we were able to evaluate both power and false positive rates using 100 replicate datasets. The results indicated that the increase in sample size from including related individuals resulted in a considerable increase in power to detect the various factors contributing to the phenotype, with little to no increase in false positives and a decrease in variability of the test statistic in the null tests (Culverhouse *et al.*, 2009).

IV. COMPUTATIONAL ISSUES

A. Permutations

Although the permutation approach implemented in the RPM software was selected for statistical reasons (see above), and permutation testing is inherently computationally expensive, the "test-by-test" construction of empirical nulls has at least two computationally appealing features.

1. Sequential evaluation

First, this approach allows for sequential evaluation of results (akin to formal sequential testing (Wald, 1945)). To illustrate, suppose you are a researcher who is performing 100 independent tests and plan to do a Bonferroni correction. As a result, you are only interested in results with p-values less than 0.001, but for anything that surpasses that threshold, you would like to have reliable estimates out to the fifth decimal place. In order to get an accurate estimate for the fifth decimal place, one would typically perform at least 10^6 permutations. After doing this once, you notice that most of the p-values are something like 0.713585 and realize that for those, you really do not care about the fifth decimal place—the 0.7 was good enough. Overall, you notice that over 99% of your permutations were performed to derive extra decimal places for results that are not of interest. Doing a few calculations, you notice that the chances that one of your "interesting" results (test statistic with an empirical p-value <0.001) would have had been surpassed by at least 3 of the first 10 permutation-based test statistics is approximately 1.2×10^{-7}. The chance that it would have been "beaten" by four or more permutation-based test statistics is less than 10^{-9}.

The next time you do a similar experiment, you "peek" at the results after 10 permutations. For any test where the unpermuted test statistic is less than four or more of the permutation based statistics, you decide not to waste your time performing any more permutations, call it "not statistically significant" and move on to the next test.

This is what the RPM software does. It "peeks" at permutation results after 10, 100, 1000, 10,000 and 100,000 permutations. The user gets to choose (by modifying the parameter file) what cutoffs to use to halt permutations earlier than the maximum number. (Of course, since the empirical p-value must equal 1 whenever the algorithm does not find any groups in the data (i.e., the test statistic $(R^2)=0$), the software never does any permutations for those tests.) In a dataset where associations to the phenotype are expected to be sparse, this sequential approach can reduce the time for computation dramatically.

2. Parallel computation

The second computational benefit of this permutation strategy is that it makes each test independent of the others. As a result, RPM analyses can easily distributed to multiple processors or computers, again potentially resulting in a considerable reduction in computation time.

B. Is the RPM too computationally expensive for modern datasets?

Clearly, any approach that exhaustively examines all pairs of SNPs (or higher order models) in a dataset with many SNPs will be computationally expensive. The use of permutation testing to evaluate statistical significance only exacerbates the problem. So the question naturally arises, is this approach practical for current genome wide genetic data?

This issue was also investigated as part of the GAW 16 (described above). Recall that this study involved a dataset of 42,461 real SNPs genotyped on 6479 family based subjects from the FHS matched to multiple simulated phenotypes. Because the phenotypes were simulated and the true model was known, distributions of false positives multiple phenotypes based on analyses of the full data and on a subset of unrelated individuals were estimated by sampling 4×10^8 SNP pairs across 100 replicate datasets, with p-values estimated by up to 10^7 permutations per pair. This constituted approximately 44% of the computational effort required to fully analyze a single dataset of this size. The analysis took approximately 3 weeks on a Macintosh Pro, containing two 3 GHz Intel Xeon processors and 16 GB of RAM. A similar analysis on data with 1 million genotyped SNPs would be expected to take approximately 1250 times as long. Because the algorithm and software lend themselves to distributed processing, larger analyses can easily be split among multiple computers. Thus, while such an analysis would be impractical for researchers limited to a single desktop computer, it would be possible for researchers with multiple computers or a computer cluster.

V. OTHER PRACTICAL ISSUES

A. Carefully clean your data before analysis

The computational burden of joint analyses in general, and of the RPM in particular, can be great. When performing univariate analyses on 1000 SNPs, including an additional SNP of poor quality adds just one more test of dubious validity. In a pairwise analysis, it adds 1000 tests, and in a 3-way analysis, it adds approximately 500,000 tests. In general, the number of possible SNP pairs grows at the square of the number of SNPs. Thus, a 10% reduction in SNPs will save nearly 20% of the tests performed. A 50% reduction in the number of SNPs will decrease the number of tests by 75%. Although we do not want to miss anything important, before an RPM analysis, certain SNPs can and should be removed from the data.

1. Remove nonpolymorphic SNPs

Always check your data for nonpolymorphic SNPs, and remove them.

2. Remove redundant SNPs

If you have a pair of SNPs with complete LD $r^2 = 1$ (i.e., they are completely redundant) only retain one for your analysis. You should check this in your own data.

If you decide to thin further based on LD, pick a high LD threshold (e.g., $r^2 > 0.9$, 0.95, or 0.99). Although you do not want to waste time and space dealing with redundant information, joint effects can be more subtle than main effects. Our experience indicates that using a threshold of 0.8 for thinning could cause you to overlook important results.

3. Remove unreliable SNPs

In current GWAS genotyping, it is common to require at least a 98% call rate for SNP to be included. Not all genotyping techniques can match this, but there should not be substantial missing data for any SNP in your analysis. Our experience is that SNPs with considerable missing data are associated with false positive signals. Thus, including such SNPs in an analysis increases both errors and computation time.

B. Additional strategies to save computation time

1. Thin SNPs further

The three cleaning approaches mentioned above for eliminating SNPs should always be applied. However, if the data still contain too many SNPs to for your computational resources, you might consider prioritizing your SNPs and only analyzing a more limited set of top candidate SNPs. Ways to rank your SNPs *a priori* include as data-driven approaches (e.g., variants on the statistical method ReliefF (Greene *et al.*, 2009)) and candidate gene approaches (which can include such information as evolutionary conservation, predicted or known function, and previous reported associations in the research literature).

2. Set a threshold for the test statistic before performing permutations

Because the null distributions are derived on a test-by-test basis, it is possible to remove tests from the analysis before permutations with no loss to the individual empirical nulls. Thus, if you decide that you are not interested in pursuing joint models that account for less than 0.1% of the trait variance in your data, you may choose not compute *p*-values for those results. By only computing *p*-values for signals of "clinical" interest, much of the computational burden of an analysis can be eliminated.

C. Statistical issues

1. Minimum cell count

Particularly when analyzing a quantitative trait, the possibility of outliers is of concern. In extreme cases, outliers with rare multilocus genotypes can result in inflated estimates of explained variance. To avoid this problem, the RPM software allows one to set the minimum cell count for which the explained variance will be computed. The minimum cell count for which a variance can be computed is 2, and the software will allow this. However, we recommend that this only be used to obtain a feel of the distribution of your data. Using a minimum cell count of 5 or 10, when a dataset has a large enough sample to support it, would be preferred.

2. Large datasets

As samples get larger, even small mean differences can become highly significant. As a result, if you have a particularly large sample, you may want to use a *p*-value threshold lower than 0.05 in order for two genotype cells to be called "distinct" (and thus not be merged by the algorithm). This can be done in the current software by setting a *p*-value for merging in the parameter file.

3. Covariates

When analyzing a quantitative phenotype using the RPM, the covariate adjustment can be addressed by regression; that is, perform a regression on the trait using the covariates and save the residuals. The residuals become the covariate adjusted trait to be analyzed using the RPM.

Covariate adjustment for dichotomous traits is a challenge. Discrete traits can be included in the analysis as predictors, and quantitative traits can be discretized using thresholds. The drawbacks to this approach are (i) there is likely to be a loss of information when a quantitative trait is discretized, and (ii) including extra predictors adds new levels of stratification. This can quickly result in sparse cells, which in turn can exacerbate the Large *P*, Small *N* problem. When data becomes sparse, typically because there are more predictors (*P*) than there are observations (*N*), models that actually arise due to random chance can fit the data extremely well. In general, one should try to keep the observations per cell as large as possible, ideally over 10.

D. Software availability

The RPM software is written in C. It is free and open source under the Gnu General Public License. Please feel free to modify the code. To obtain the software and user manual, please write to rculverh@wustl.edu.

VI. SUMMARY

The RPM is an exploratory tool to examine the joint effect of multiple genetic loci (and environmental exposures) on phenotypic variation. The method is designed to be sensitive to factors whose effects may be expressed primarily as interactions, displaying small marginal effects; but it is not a test of interactions. The method is appropriate for either quantitative or dichotomous traits. The computational challenge of examining all pairs, or higher-order joint effects, is ameliorated by the fact that the analyses can easily be distributed to multiple processors.

Acknowledgments

This work was supported by National Institutes of Health grants K25 GM69590 from the National Institute of General Medical Sciences, R03 DA023166 from the National Institute on Drug Abuse (NIDA), and CA89392 from the National Cancer Institute, as well as IRG-58-010-50 from the American Cancer Society. There are no known conflicts of interest.

References

Anholt, R. R., Dilda, C. L., Chang, S., Fanara, J. J., Kulkarni, N. H., Ganguly, I., Rollmann, S. M., Kamdar, K. P., and Mackay, T. F. (2003). The genetic architecture of odor-guided behavior in Drosophila: Epistasis and the transcriptome. *Nat. Genet.* **35**(2), 180–184.

Churchill, G. A., Airey, D. C., Allayee, H., Angel, J. M., Attie, A. D., Beatty, J., Beavis, W. D., Belknap, J. K., Bennett, B., Berrettini, W., *et al.* (2004). The Collaborative Cross, a community resource for the genetic analysis of complex traits. *Nat. Genet.* **36**(11), 1133–1137.

Comtet, L. (1974). Advanced Combinatorics: The Art of Finite and Infinite Expansions D. Reidel Pub. Co, Dordrecht, Boston.

Culverhouse, R. (2007). The use of the restricted partition method with case-control data. *Hum. Hered.* **63**(2), 93–100.

Culverhouse, R., Suarez, B. K., Lin, J., and Reich, T. (2002). A perspective on epistasis: Limits of models displaying no main effect. *Am. J. Hum. Genet.* **70**(2), 461–471.

Culverhouse, R., Klein, T., and Shannon, W. (2004). Detecting epistatic interactions contributing to quantitative traits. *Genet. Epidemiol.* **27**(2), 141–152.

Culverhouse, R., Jin, W., Jin, C. H., Hinrichs, A. L., and Suarez, B. K. (2009). Power and false-positive rates for the restricted partition method (RPM) in a large candidate gene data set. *BMC Proc.* **3**(Suppl. 7), S74.

Cupples, L. A., Heard-Costa, N., Lee, M., and Atwood, L. D. (2009). Genetics Analysis Workshop 16 Problem 2: The Framingham Heart Study data. *BMC Proc.* **3**(Suppl. 7), S3.

Edwards, T. L., Turner, S. D., Torstenson, E. S., Dudek, S. M., Martin, E. R., and Ritchie, M. D. (2010). A general framework for formal tests of interaction after exhaustive search methods with applications to MDR and MDR-PDT. *PLoS ONE* **5**(2), e9363.

Galvan, A., Ioannidis, J. P., and Dragani, T. A. (2010). Beyond genome-wide association studies: Genetic heterogeneity and individual predisposition to cancer. *Trends Genet.* **26**(3), 132–141.

Games, P. A., and Howell, J. F. (1976). Pairwise multiple comparison procedures with unequal N's and/or variances: A monte carlo study. *J. Educ. Stat.* **1**, 113–125.

Gerke, J., Lorenz, K., and Cohen, B. (2009). Genetic interactions between transcription factors cause natural variation in yeast. *Science* **323**(5913), 498–501.

Greene, C. S., Penrod, N. M., Kiralis, J., and Moore, J. H. (2009). Spatially Uniform ReliefF (SURF) for computationally-efficient filtering of gene–gene interactions. *BioData Min* **2**(1), 5.

Hahn, L. W., Ritchie, M. D., and Moore, J. H. (2003). Multifactor dimensionality reduction software for detecting gene–gene and gene–environment interactions. *Bioinformatics* **19**(3), 376–382.

Hirschhorn, J. N. (2009). Genomewide association studies—Illuminating biologic pathways. *N. Engl. J. Med.* **360**(17), 1699–1701.

Kraft, P., and Hunter, D. J. (2009). Genetic risk prediction—Are we there yet? *N. Engl. J. Med.* **360**(17), 1701–1703.

Kraja, A. T., Culverhouse, R., Daw, E. W., Wu, J., Van Brunt, A., Province, M. A., and Borecki, I. B. (2009). The Genetic Analysis Workshop 16 Problem 3: Simulation of heritable longitudinal cardiovascular phenotypes based on actual genome-wide single-nucleotide polymorphisms in the Framingham Heart Study. *BMC Proc.* **3**(Suppl 7), S4.

Lander, E. S., and Schork, N. J. (1994). Genetic dissection of complex traits. *Science* **265**(5181), 2037–2048.

Mackay, T. F. (2010). Mutations and quantitative genetic variation: Lessons from Drosophila. *Philos. Trans. R. Soc. Lond. B Biol. Sci.* **365**(1544), 1229–1239.

Manolio, T. A., Collins, F. S., Cox, N. J., Goldstein, D. B., Hindorff, L. A., Hunter, D. J., McCarthy, M. I., Ramos, E. M., Cardon, L. R., Chakravarti, A., *et al.* (2009). Finding the missing heritability of complex diseases. *Nature* **461**(7265), 747–753.

Nelson, M. R., Kardia, S. L., Ferrell, R. E., and Sing, C. F. (2001). A combinatorial partitioning method to identify multilocus genotypic partitions that predict quantitative trait variation. *Genome Res.* **11**(3), 458–470.

Pedersen, J. C., and Berg, K. (1989). Interaction between low density lipoprotein receptor (LDLR) and apolipoprotein E (apoE) alleles contributes to normal variation in lipid level. *Clin. Genet.* **35**(5), 331–337.

Pedersen, J. C., and Berg, K. (1990). Gene–gene interaction between the low density lipoprotein receptor and apolipoprotein E loci affects lipid levels. *Clin. Genet.* **38**(4), 287–294.

Phillips, P. C. (2008). Epistasis—The essential role of gene interactions in the structure and evolution of genetic systems. *Nat. Rev. Genet.* **9**(11), 855–867.

Risch, N., and Merikangas, K. (1996). The future of genetic studies of complex human diseases. *Science* **273**(5281), 1516–1517.

Routman, E. J., and Cheverud, J. M. (1995). Gene effects on a quantitative trait: Two-locus epistatic effects measured at microsatellite markers and at estimated QTL. *Evolution* **51**, 1654–1662.

Sanjuan, R., and Elena, S. F. (2006). Epistasis correlates to genomic complexity. *Proc. Natl. Acad. Sci. USA* **103**(39), 14402–14405.

Schork, N. J. (1997). Genetics of complex disease: Approaches, problems, and solutions. *Am. J. Respir. Crit. Care Med.* **156**(4 Pt 2), S103–S109.

Sing, C. F., Stengard, J. H., and Kardia, S. L. (2003). Genes, environment, and cardiovascular disease. *Arterioscler. Thromb. Vasc. Biol.* **23**(7), 1190–1196.

Szathmary, E., Jordan, F., and Pal, C. (2001). Molecular biology and evolution. Can genes explain biological complexity? *Science* **292**(5520), 1315–1316.

Templeton, A. (2000). Epistasis and complex traits. *In* "Epistasis and the Evolutionary Process" (J. B. Wolf, E. D. Brodie, and M. J. Wade, eds.), pp. 41–57. Oxford University Press, Oxford.

Vieira, C., Pasyukova, E. G., Zeng, Z. B., Hackett, J. B., Lyman, R. F., and Mackay, T. F. (2000). Genotype–environment interaction for quantitative trait loci affecting life span in *Drosophila melanogaster*. *Genetics* **154**(1), 213–227.

Wald, A. (1945). Sequential tests of statistical hypotheses. *Ann. Math. Stat.* **16**(2), 117–186.

Wolf, J. B., Leamy, L. J., Routman, E. J., and Cheverud, J. M. (2005). Epistatic pleiotropy and the genetic architecture of covariation within early and late-developing skull trait complexes in mice. *Genetics* **171**(2), 683–694.

7 Statistical Methods for Pathway Analysis of Genome-Wide Data for Association with Complex Genetic Traits

Peter Holmans

Biostatistics and Bioinformatics Unit, MRC Centre for Neuropsychiatric Genetics and Genomics, Department of Psychological Medicine and Neurology, Cardiff University School of Medicine, Heath Park, Cardiff, United Kingdom

Advances in Genetics, Vol. 72

0065-2660/10 $35.00
DOI: 10.1016/S0065-2660(10)72007-1

ABSTRACT

A number of statistical methods have been developed to test for associations between pathways (collections of genes related biologically) and complex genetic traits. Pathway analysis methods were originally developed for analyzing gene expression data, but recently methods have been developed to perform pathway analysis on genome-wide association study (GWAS) data. The purpose of this review is to give an overview of these methods, enabling the reader to gain an understanding of what pathway analysis involves, and to select the method most suited to their purposes. This review describes the various types of statistical methods for pathway analysis, detailing the strengths and weaknesses of each. Factors influencing the power of pathway analyses, such as gene coverage and choice of pathways to analyze, are discussed, as well as various unresolved statistical issues. Finally, a list of computer programs for performing pathway analysis on genome-wide association data is provided.

I. INTRODUCTION

For most complex traits, the underlying biology is not well characterized, making it difficult to select candidate genes for analysis. In recent years, therefore, researchers have concentrated on performing genome-wide genetic studies to determine disease susceptibility genes for such traits. Initially, genome-wide linkage studies were popular, but currently gene expression studies based on microarrays and genome-wide association studies (GWAS), based on SNP arrays, are more common. The first-pass analysis of such studies typically tests the relationship of each individual gene (for microarray or proteomic studies) or SNP (for GWAS) with disease separately. Such approaches can be quite successful—for example, several highly significant SNP-disease associations have been found through GWAS that have replicated across studies (Manolio *et al.*, 2008). However, the effect sizes for these associations have generally been quite small, and do not explain the observed heritability of most traits (Maher, 2008). Recently, therefore, interest has grown in testing association of sets of genes or

SNPs with disease simultaneously. Typically, these genes are related to each other biologically, falling within a biological pathway, and so these analyses of gene sets are often called "pathway" analyses.

Intuitively, it seems likely that susceptibility alleles for any given disorder are not randomly distributed among genes but instead, are distributed among one (or more) set(s) of genes whose functions are to some extent related. Compared with single-locus analysis, group or pathway analysis may yield more secure insights into disease biology since an associated pathway is likely to implicate function better than a hit in a single gene that may have many functional possibilities. Additionally, genetic heterogeneity may cause any one causal variant to exhibit only modest disease risk in the sample as a whole, since different individuals may possess different disease risk alleles at different loci in the same gene, or in different genes. This will reduce power to detect any one variant by traditional association methods. However, if the genes in question are members of the same biological pathway, then considering the pathway as the unit of analysis may increase power to detect association between the genes and disease. For similar reasons, association of disease with biological pathways may be easier to replicate across different studies than association to individual SNPs or genes. This was clearly shown in an analysis of Crohn's disease (Wang *et al.*, 2009), where the IL12/IL23 pathway showed evidence of enrichment in four independent GWAS, despite the genes and SNPs involved differing between the studies. Pathway analysis approaches can thus be regarded as complementary to the studies that focus on the top hits. In gene expression analysis, pathway-based approaches have been shown to yield significant results even in the absence of single-gene associations (e.g., Mootha *et al.*, 2003). There have also been a number of promising disease associations reported recently from pathway analyses of GWAS data (e.g., Askland *et al.*, 2009; O'Dushlaine *et al.*, 2010; Wang *et al.*, 2009). Although pathway analyses of complex genetic traits show considerable potential for elucidating disease biology and finding novel disease susceptibility genes, several statistical issues remain unresolved. Particularly for GWAS data (Cantor *et al.*, 2010; Kraft and Raychaudhuri, 2009). In this chapter, the various statistical methods for pathway analysis are reviewed, particular attention being paid to the issues raised by GWAS data, and their relative advantages and disadvantages discussed. A brief overview of publically available pathway databases is given. Unresolved statistical issues are discussed, and a list of software packages for performing pathway analysis on genome-wide association data given.

II. WHAT IS PATHWAY ANALYSIS?

In pathway analysis, a set of genes (the "pathway") is tested for enrichment of association signal with a trait. That is, do the genes show significantly more association than expected under the null hypothesis. Pathway analysis methods fall into two types, according to the null hypothesis being tested (Goeman and Bühlmann, 2007): *Competitive* tests compare the association between a gene-set and disease with that of all other gene sets being studied. In other words, the null hypothesis being tested is that the association between gene-set and disease is *average*, given the overall distribution of association across the genome. Conversely, *self-contained* tests test whether there is significant association between the gene-set and disease. Here, the null hypothesis being tested is that there is *no* association between the gene-set and disease. While the latter may seem more natural, the former have the advantage of being robust to systematic inflation of association statistics across the genome (Devlin and Roeder, 1999), whether this is due to population stratification (Price *et al.*, 2006), differential rates of genotyping error between cases and controls (Moskvina *et al.*, 2006), or other reasons, provided the inflation across genes in the gene-set is similar to that in the rest of the genome.

III. STATISTICAL CONSIDERATIONS WHEN ANALYZING GWAS DATA

Many pathway methods test for enrichment of association signal by comparing the significance of genes within the gene-set with a background distribution (often the genes covered by the genotyping/expression chip which are not part of the gene-set). This requires a single association test statistic to be defined for each gene. For gene expression studies, there is typically only one observation per gene, making the choice of test statistic simple. For GWAS data, however, there are often several SNPs per gene, with various possibilities for defining gene-wide significance (these are discussed in more detail later). When assessing gene-wide significance, it is important to bear in mind the varying numbers of SNPs per gene. For example, many methods define the measure of gene-wide significance to be the most significant single-SNP p-value, or largest single-SNP association statistic. This measure will depend on gene size, since large genes containing many SNPs are more likely to have significant SNPs by chance than genes with few SNPs. Thus, pathway analyses that do not correct for gene size will favor pathways containing large genes. This relationship between unadjusted minimum p-value and gene size was confirmed by Hong *et al.* (2009), who also noted that several biologically relevant classifications, such as synapse, cell adhesion, and nervous system development, were enriched for large genes. Thus, when pathway analyses of GWAS data report "biologically meaningful" results, it is important to check whether the analyses correct for gene size. The association statistics for SNPs that are close together will not be independent, due to linkage disequilibrium (LD). Typically, LD rarely extends beyond a few hundred kilobases (Weiss and Clark, 2002), but larger regions do exist, such as the MHC region on chromosome 6. This nonindependence means that standard techniques for correcting the most significant *p*-value for the number of SNPs, such as the Bonferroni method, will not be valid. Furthermore, LD raises the possibility that significant *p*-values in genes that are close together may reflect the same association signal. If these genes are members of the same pathway, this could lead to false evidence for association to that pathway. Hong *et al.* (2009) found that genes did indeed cluster nonrandomly with respect to biological annotation, so this is a potentially serious problem. Genes sometimes overlap physically, resulting in some SNPs lying within more than one gene. This raises the problem of how to assign SNPs to genes. If a SNP is assigned to all genes in which it lies, and more than one of these lie in a particular pathway, this could give false evidence of enrichment. Conversely, if the SNP is to be assigned to only one gene, the question arises of how this assignment is to be performed. If individual genotype data are available, correction can be made for LD, gene size and overlapping genes by randomly permuting the phenotype data among individuals and repeating the GWAS a large number of times. These replicate GWAS studies, generated under the assumption of random association of genes (and thus pathways) to phenotype provide a distribution against which the pathway association statistics from the actual data can be compared. While such permutation analyses are appropriate statistically, they rely on individual genotype data being available. Furthermore, generating a sufficiently large number of replicate GWAS can be time consuming and computationally intensive, particularly for large samples genotyped on large numbers of SNPs. Therefore, some methods have been developed which do not require whole-genome permutations to be performed.

IV. STATISTICAL METHODS FOR PATHWAY ANALYSIS

In this section, the various different types of pathway analysis methods are briefly reviewed, along with their advantages and disadvantages.

A. Overrepresentation analysis

In an overrepresentation analysis, a list of significant genes is obtained from the observed data by selecting those genes whose gene-wide significance reaches a predetermined threshold. The proportion of genes in a pathway appearing on the list is compared with the proportion of genes not in the pathway appearing on the list, and a statistical test performed to test for differences in these proportions. If the proportion of genes in the pathway appearing on the list is significantly higher than the corresponding proportion of genes not in the pathway, the pathway is said to be overrepresented. Overrepresentation analyses are thus examples of competitive tests. This approach has been implemented in several analysis packages for microarray data, for example, EASE (Hosack *et al.*, 2003);

GOStat (Beissbarth and Speed, 2004); DAVID (Dennis *et al.*, 2003; Huang da *et al.*, 2009a) and MetaCore™, part of the GeneGO suite of systems biology analysis programs (Bugrim *et al.*, 2004). Most of these packages base their statistical tests for overrepresentation on either Fisher's exact test, the hypergeometric distribution, or a chi-square analysis of the 2×2 contingency table (Fig. 7.1). All of these tests should give similar results if large lists of significant genes and pathways are used; the first two are more accurate for small gene lists and pathways. All of these tests assume that the presence of a gene on the list of significant genes does not depend on the other genes on the list. This may not be true, due to correlated expression among genes, which may result in inflated significance of the overrepresentation test. The choice of genes to which the pathway genes are compared is important. Some packages, such as GOStat, use all genes in the genome regardless of whether they are covered by the array used to generate the data, or whether they have relevant biological annotations (i.e., annotated to at least one of the pathways being tested). Some packages, such as DAVID, use only annotated genes that are on the array. This is more conservative, but probably more accurate. Other packages allow the user to upload a custom reference list (e.g., removing the genes that failed quality control). While the choice of the genes used is likely to affect the absolute *p*-values for overrepresentation for each pathway, it will have less effect on the ranking of the pathways in order of significance (Huang da *et al.*, 2009b).

	Genes on list	Genes not on list	
Genes in pathway	a	b	a+b
Genes not in pathway	c	d	c+d
	a+c	b+d	

Figure 7.1. The idea of overrepresentation analysis is to test whether the number of genes in the pathway on the list of significant genes (a) is significantly higher than would be expected given the total number of genes in the pathway (a+b), the total number of genes not in the pathway (c+d), the number of genes on the list (a+c) and the number of genes not on the list (b+d).

For a more extensive review of the various methods for overrepresentation analysis of microarray data, the statistical methods they use, and other analysis issues, see Khatri and Drăghici (2005) and Huang da *et al.* (2009b).

One of the first overrepresentation analyses of GWAS data was performed by Torkamani *et al.* (2008) on the Wellcome Trust Case Control Consortium (2007) data. They assigned SNPs to genes based on a 5 kb window either side of the gene. Each SNP was assigned to one gene only (priorities based on location relative to the gene). Genes were ranked according to their most significant *p*-value (no correction for multiple SNPs). The top 2.5% of genes were tested for overrepresentation using MetaCore™.

Askland *et al.* (2009) analyzed the Wellcome Trust Case Control bipolar disorder data using their software package exploratory visual analysis (EVA). Their analysis assigned SNPs to the nearest gene (regardless of distance) from Affymetrix 500K array annotation files. Approximately 25% of SNPs were assigned to more than one gene in these files, and these were discarded from the analysis. The significance measure for each gene was defined as the most significant single-SNP *p*-value, and an overrepresentation analysis was performed on the top 10% of genes. Two statistical tests were used to perform the overrepresentation analysis: Fisher's exact test, and a permutation method based on randomly permuting gene labels among genes (thus, assuming that significance levels of genes are independent of each other). No correction for gene size was applied. Pathways related to ion channels were found to be overrepresented, which is biologically plausible (Ferreira *et al.*, 2008). However, it should be noted that several of the ion channel genes are large, with many SNPs.

Holmans *et al.* (2009) attempted to correct for variable numbers of SNPs in genes by taking the list of SNPs with association *p*-value less than some predefined criterion, and defining the list of significant genes as those genes to which the SNPs are assigned. If a SNP lay within more than one gene, it was assigned to all genes. Each gene is counted only once, regardless of how many SNPs on the list are assigned to that gene, thus removing the problem of correcting for LD between multiple SNPs in a gene. Several thousand replicate gene lists of the same length are generated by sampling SNPs at random from the entire set of SNPs in the study, and adding their genes to the list. This corrects for variable numbers of SNPs per gene, since large genes with many SNPs are more likely to be sampled on the replicate gene lists. The number of genes in each pathway on the gene list is counted, and compared with that on each of the replicate gene lists to obtain a *p*-value for overrepresentation. Correction of these *p*-values for testing multiple nonindependent pathways was done by a bootstrap method: one of the replicate gene lists was selected at random as the "real study," and was tested against a sample of gene lists selected at random (with replacement) from the remaining replicate gene lists, thus obtaining overrepresentation *p*-values for each pathway. This procedure is repeated many times, and the corrected p-value for each pathway in the real data defined as the proportion of simulated "studies" containing a pathway at least as significant. This procedure also provides a test of whether the number of significantly overrepresented pathways in the real data is higher than expected by chance (allowing for nonindependence of pathways). The simulation procedure used in this method is considerably faster than permuting phenotypes and allows for SNPs lying in several overlapping genes (although such a SNP will add all its genes to the gene list, it is only allowed to add one to the score for each pathway, to correct for situations where there are families of genes with similar biological annotations at the same locus). Furthermore, it does not require individual genotype data to be available. However, it does not allow for LD between genes, such that significant SNPs in genes that are close together may be reflecting the same association signal. The authors recommend the removal of genes from regions showing long-range LD (e.g., the MHC region on chromosome 6) if an association signal is observed there, although such an approach is conservative. The method is implemented in a computer program, ALIGATOR.

A different approach to correcting for LD between SNPs was adopted by Hong *et al.* (2009). They clustered SNPs that were highly correlated ($r^2 > 0.8$) together. Then, they defined gene-wide significance by the minimum association p-value for any SNP assigned to the gene multiplied by the number of clusters overlapping the gene. A correction for LD between genes was applied by removing genes whose most significant SNP lay within a cluster which contained more significant SNPs assigned to other genes. SNPs were assigned to genes if they lay within 1 kb of the gene. The gene-wide significance levels thus obtained were then used to produce gene lists for analysis using packages such as DAVID or GSEA (Subramanian *et al.*, 2005). Removing genes due to LD with more significant genes is an appealing idea, but may cause problems for analyses based on ranking of genes, such as GSEA (Hong *et al.*, 2009). Additionally, it may not always be clear which gene is driving an association signal, so removing genes based on LD may reduce power to detect enriched pathways (if the gene removed is a member of the pathway but the gene retained is not). Note that the correction for multiple SNPs in a gene used to obtain the gene-wide significance could result in the list of significant genes not necessarily containing those with the most significant single-SNP p-values (unlike the correction used by Holmans *et al.*) A program for producing the gene lists, ProxyGeneLD, is available from the authors.

Rather than defining their gene list as the genes with the most significant gene-wide p-values, Baranzini *et al.* (2009) used protein interaction networks to define modules of genes for further testing. Analysis was restricted to genes containing a SNP with p-value < 0.05 (no correction for number of SNPs per gene), and modules were generated using a package called jActive modules (Ideker *et al.*, 2002). The method grows a module from each node by adding a neighboring gene from the interaction network and computing an aggregate score for association $S = \sqrt{(kZ)}$, where the p-value for each gene is converted into a Z-score, and k is the number of genes in the module. Once S stops growing, the module is regarded as complete, and its significance tested against the association scores from modules randomly selected from the network. Biological meaning of the resulting modules was assessed by testing for overrepresentation of prespecified pathways on their lists of genes. When applied to multiple sclerosis GWAS data, the method implicated neural pathways in MS susceptibility, a novel finding. However, it should be noted that several of the genes involved are large, so the results may be due in part to the lack of correction for gene size.

All of the methods for analyzing GWAS data described thus far have used the best single-SNP p-value to define the gene-wide significance measure (with or without a correction for gene size). While this is certainly a sensible measure to use, it does not allow for the presence of multiple independent association signals in a gene. If a gene, or pathway, contains several rare mutations, it is likely to have several moderately associated SNPs, and such association may be missed by methods that concentrate solely on the most significant SNPs. O'Dushlaine *et al.* (2010) developed the SNP-ratio test (SRT) to deal with such situations. In this method, the proportion of significant SNPs in a pathway (those with association p-value <0.05) in the real data is compared with that in simulated datasets obtained by permuting phenotypes among individuals and repeating the GWAS. The list of significant SNPs in the simulated datasets is defined as the top N SNPs, where N SNPs had $p < 0.05$ in the real data, rather than all SNPs with $p < 0.05$. The test is thus robust to genome-wide inflation of association test statistics (due to population stratification, for example), like other competitive tests. Unlike the other methods described in this section, the SRT requires individual genotype data to be available and, since it uses whole-genome permutation analyses, is computationally intensive. It does not use the actual p-values of the SNPs, which may cause it to lose power compared to set-based analysis methods.

B. Gene-set enrichment methods

A major problem of the overrepresentation methods discussed in the previous section is that they all require a threshold to be specified in order to define the list of significant genes or SNPs. It is seldom clear what the optimal choice of threshold should be, and different thresholds might yield different overrepresented pathways. Gene-set enrichment analyses evade the problem of defining a threshold by instead ranking the genes in order of significance, and then testing for differences between the ranks of genes in a pathway compared to other genes. Thus, gene-set enrichment analyses are examples of competitive tests.

The original gene-set enrichment analysis (GSEA) method (Subramanian *et al.*, 2005) was designed to analyze gene expression data, and works as follows:

1. Rank order the N genes in the dataset ($g_1, g_2, \ldots, g_N$) in order of their test statistic r_j.
2. For a given gene-set (pathway) G, for each j ($1 \leq j \leq N$), calculate the quantity

$$E(G,j) = (A/B) - (C/D),$$

   where A = sum of test statistics $|r_i|^p$ for all genes in G with rank $\leq j$; B = sum of test statistics $|r_i|^p$ for all genes in G; C = number of genes not in G with rank $\leq j$; D = total number of genes not in G.
3. The maximum deviation of the $E(G,j)$ from zero is denoted the *enrichment score* for gene-set G: $ES(G)$, and measures whether the distribution of ranks among genes in G is nonrandom. If genes in G tend to have higher than expected rank, $ES(G)$ will be positive, whereas if they rank lower than expected, ES(G) will be negative. Note that if $p=0$, $ES(G)$ is the standard Kolmogorov–Smirnov statistic. Subramanian *et al.* recommended using $p=1$, thereby giving higher weight to genes with larger test statistics.
4. Randomly permute the phenotypes among individuals and repeat the gene expression analysis. Rank the genes in order of significance, and denote the enrichment score for gene-set G in permutation π by $ES(G,\pi)$. This process is repeated 1000 times, and the empirical significance of the observed enrichment score $ES(G)$ is the proportion of $ES(G, \pi) < ES(G)$.

The method is implemented in a software package, GSEA-P. A generalization of the method was presented by Barry *et al.* (2005), in the SAFE analysis package, available through Bioconductor (http://www.bioconductor.org), a collection of open source software for bioinformatics. In SAFE, the user is able to specify gene-specific statistics ("local statistics"), and ways of combining statistics across genes in a pathway ("global" statistics). The GSEA enrichment score is an example of a "global" statistic. Significance of both local and global statistics is assessed by permuting phenotypes among individuals and recomputing the statistics. Both GSEA-P and SAFE are notable as the only methods for testing pathway enrichment in microarray data that use permutation of phenotypes among individuals, thus correctly accounting for correlation between expression values of related genes (Allison *et al.*, 2006).

The first extension of GSEA to GWAS data was proposed by Wang *et al.* (2007). Two gene-wide test statistics were considered: the minimum single-SNP p-value (uncorrected for numbers of SNPs in a gene) and the Simes correction (Simes, 1986). The latter method is defined as the minimum of $\{p_{(i)}L/i\}$, where $p_{(i)}$ is the ith most significant single-SNP p-value in the gene,

L is the number of SNPs in the gene, and $1 \leq i \leq L$. The latter approach is probably overconservative (Wang *et al.*, 2007), and also results in genes no longer being ranked in order of their most significant SNP, which may be undesirable. Conversely, applying no correction leads to bias in favor of large genes (Wang *et al.*, 2007). SNPs were assigned to genes if they lay within that gene, otherwise to the nearest gene (up to 500 kb away). Significance of the enrichment score for each pathway was assessed by randomly permuting the gene-wide statistics among genes, although this does not allow for overlapping genes (with SNPs common to both), or LD between genes. A later implementation of the method (Wang *et al.*, 2009), implemented in the software package GenGen, uses the most significant SNP as the gene-wide statistic, and calculates significance of enrichment scores by permuting phenotypes among individuals and repeating the genome-wide association analysis. This approach corrects for LD between SNPs, but is computationally intensive. A version of GSEA based on individual SNPs, rather than genes, was developed by Holden *et al.* (2008), and implemented in a software package GSEA-SNP. This method also calculates significance of enrichment scores in this method by permuting phenotypes among individuals and repeating the genome-wide association analysis.

Compared to overrepresentation analyses, gene-set enrichment methods have two advantages. First, they do not require an arbitrary threshold to be specified in order to define a list of "significant" genes. Second, they make use of the actual significance of the associations of the genes with the trait being analyzed, giving more weight to more significantly associated genes. This may potentially increase power to detect significantly enriched pathways. However, it is possible that a single highly associated gene may on its own give a high enrichment score to any pathway of which it is a member (Hong *et al.*, 2009; Wang *et al.*, 2009). Thus, a high enrichment score for such a pathway would not necessarily reflect involvement of the pathway as a whole with disease. Thus, care must be taken applying gene-set enrichment methods to datasets containing one or more highly significantly associated genes (or SNPs)—perhaps the gene-set enrichment analysis should be repeated with such genes removed.

C. Set-based methods

Set-based methods aggregate the association evidence across all genes (or SNPs) in a pathway into one combined test statistic, and then test whether this statistic is larger than would be expected under the null hypothesis that there is no association between any gene in the pathway and the phenotype. Thus, set-based methods are examples of self-contained tests, unlike overrepresentation or gene-set enrichment methods. A number of set-based methods have been proposed to analyze gene expression data. These include multivariate tests, which test for relationships between phenotype and all gene expression values simultaneously, and methods which combine single-gene statistics. Examples of multivariate tests are Hotelling's T^2-test, a multivariate extension of the t-test (Lu *et al.*, 2005), the "globaltest" of Goeman *et al.* (2004), in which trait phenotype is regressed on all expression values simultaneously, and GlobalAncova (Hummel *et al.*, 2008), a multivariate analysis of covariance procedure. Methods for combining single-gene statistics include a weighted mean of t-tests (Tian *et al.*, 2005). Jiang and Gentleman (2007) proposed using the median or a sign test instead of the mean, since these are more robust to outliers. For a full review of set-based methods for gene expression analysis, see Ackermann and Strimmer (2009). It might be expected that multivariate analysis methods, which take into account correlations between expression measurements at different genes, might be more powerful than methods combining single-gene statistics, which do not, when such correlations exist. However, this is not necessarily always the case (Ackermann and Strimmer, 2009; Glazko and Emmert-Streib, 2009), so both types of method should be used (Glazko and Emmert-Streib, 2009).

For combining SNP-level p-values, one approach is to use the mean of the association statistics for the SNPs. This is implemented in the "set-based" tests in the statistical genetics package PLINK (Purcell *et al.*, 2007). Alternatively, Fisher's method for combining p-values (Fisher, 1932) can be applied—this calculates the sum of the natural logs of the p-values and multiplies this by -2. If there is no association between any SNP and the trait being analyzed, this should be distributed as a chi-square on $2n$ degrees of freedom (n being the number of SNPs), provided the SNPs are independent. Refinements of the Fisher method include the truncated product method (Zaykin *et al.*, 2002), which uses the product of all p-values less than a preselected threshold, and the rank truncated product (RTP) method (Dudbridge & Koeleman, 2003), which uses the product of the best N SNPs (where N is prespecified by the user)as its test statistic. These methods aim to increase the power of Fisher's method by discarding SNPs that are unlikely to be associated with the trait, thereby reducing noise. While theoretical distributions are available for these statistics, they rely on the p-values being independent. For SNPs within a gene, or pathway, this is not likely to be true, due to LD. Although it is possible to obtain a theoretical distribution for Fisher's method when the p-values are correlated (Makambi, 2003), currently no such distribution is known for the truncated product methods. Therefore, it is usual to assess significance by randomly permuting the trait phenotypes among individuals, and repeating the analysis, although faster simulation-based methods have been proposed (Seaman and Müller-Myhsok, 2005)

One major problem with the truncated product methods is that they require prespecification of either a p-value threshold or the number of SNPs to be included. Choosing the wrong truncation point can greatly reduce power. Dudbridge and Koeleman (2004) addressed this issue by proposing an adaptive rank-truncated product (ARTP) method, in which a number of truncation points are tested, with the minimum empirical *p*-value being used as the test statistic. However, this procedure requires correction for multiple nonindependent tests (the truncation points used), which potentially requires two levels of simulation. First, it is necessary to generate a large sample of simulated datasets (with random permutation of phenotypes), from which empirical *p*-values for each truncation point can be estimated for the observed data. Second, empirical *p*-values for each truncation point must be obtained for each of these simulated datasets, in order to build an empirical distribution for the most significant *p*-value. This requires a further sample of simulated datasets to be generated for each of the first set of simulated datasets. This procedure can quickly become computationally infeasible. Yu *et al.* (2009) devised a method for performing ARTP analysis which requires only one level of simulations, thereby reducing the computational burden. They also developed a method where the truncation is performed across gene-based *p*-values, with the gene-based *p*-value being obtained from the simulated datasets. Yu *et al.* used RTP and ARTP to define the summary statistics for each gene, but other measures could be used (e.g., best *p*-value). The gene-based ARTP was found to give superior power to the SNP-based ARTP when the sizes of the genes in the pathway were highly variable, and the associated SNPs in the smaller genes. In other situations, the powers of the two methods were similar. The method can also be adapted for use when truncation is based on *p*-value thresholds rather than rank threshold. A similar approach was proposed by de la Cruz *et al.* (2010), who base their thresholding on *p*-values, but allow these to depend on the ranked significance of the SNPs. They also propose to weight the contributions of SNPs to the product of *p*-values, to reflect LD (SNPs that are highly correlated due to LD receive lower weights).

Set-based analyses are an appealing framework for testing pathways for association with a trait because they utilize the actual association evidence for each gene or SNP in a natural and efficient manner. However, the simulations required to make them robust to correlations between genes/SNPs (e.g., due to LD) make them relatively computationally intensive. Thus, they tend to be applied either to pathways for which there is a prior biological interest, or to pathways highlighted from a previous analysis, rather than used for screening several pathways at once. Furthermore, unlike overrepresentation and gene-set enrichment analyses, set-based analyses are not robust to systematic genomic inflation of individual test statistics, such as are often observed in GWAS (Devlin and Roeder, 1999). Even small inflations, when aggregated over several SNPs, can cause large inflations in Type I error. Some authors have corrected association statistics by dividing by the genomic inflation factor (e.g., Moskvina *et al.*, 2009). However, this is likely to be conservative, since for some traits, the observed inflation may be due to several thousand true susceptibility genes of small effect (International Schizophrenia Consortium *et al.*, 2009). The best way of correcting set-based analyses for genomic inflation has yet to be determined.

D. Modeling methods

These methods attempt to model the relationship between trait phenotype and a set of genes/SNPs in a pathway using more sophisticated statistical techniques than those described in the previous sections. Examples include the use of high-dimensional interactions (Ritchie *et al.*, 2001), Monte-Carlo logic regression (Kooperberg and Ruczinski, 2005), and Bayesian modeling (Conti *et al.*, 2003). Lesnick *et al.* (2007) used stepwise regression to select the best-fitting model from all combinations of SNP main effects and pairwise interaction terms. As pointed out by Breitling *et al.* (2009), searching over a large set of possible models can lead to overfitting, resulting in *p*-values that are several orders of magnitude too low, so care must be taken in interpreting the results of such analyses. If experimental data on products of the pathway (e.g., lipid levels) are available, more complicated mathematical methods may be used to model the behavior of the pathway through time. This may in turn inform the design of future experiments, and suggest possible treatment interventions. For an example of this approach, see Pearson *et al.* (2009), who used differential equations to model the processes underlying lipoprotein cholesterol binding in the liver. Their models do not explicitly allow for genetic effects, but it might be possible to adapt the framework to include such effects.

Although modeling approaches have the potential to give insights into the way biological pathways act to influence disease risk that are not available from the other methods described, care must be taken with their statistical interpretation. Since such approaches are statistically complex and computationally intensive, they are best suited for application to a set of genes with known biological function, rather than as a screening approach based on whole genome data aimed at selecting pathways for further study.

E. Network-based methods

All the methods for pathway analysis previously described require that the pathway is already specified. Methods that derive pathways or gene networks from the data itself may therefore yield novel biological insights that are not available from other types of pathway analysis. Several methods exist for clustering genes based on their coexpression in microarray networks, resulting in clusters, or modules, of genes whose expression patterns are closely correlated, for example: weighted coexpression network analysis (Zhang and Horvath, 2005). The properties of a number of clustering methods were reviewed by Richards *et al.* (2008). Modules produced by such analysis methods were found to be highly reproducible across brain regions in human data (Oldham *et al.*, 2008). Clustering analysis can be performed across tissue types, such as hypothalamus, liver, and adipose tissue (Dobrin *et al.*, 2009), or in nonhuman data, such as yeast (Zhu *et al.*, 2008). In order to validate the biological properties of the modules, their genes are often tested for enrichment for genes in known biological pathways, such as gene ontology (GO) categories (Ashburner *et al.*, 2000; Harris *et al.*, 2004), cell types (Lein *et al.*, 2007), or genes associated with clinical traits. Gene modules from clustering analyses can sometimes be quite large (Richards *et al.*, 2008), and thus heterogeneous with respect to their biological characteristics. Module membership was proposed by Oldham *et al.* (2008) as a means of quantifying how closely each gene in a module conforms to the characteristic expression pattern of that module, and thus identify the genes most likely to participate in the biological functions represented by that module. Module membership is defined as the correlation between the expression pattern of that gene and the first principal component of the expression data for all genes in the module, and was found to correlate with marks of cell type. Methods have also been proposed for examining the relationship between gene networks and disease susceptibility using other types of data, for example: protein–protein interactions (Lage *et al.*, 2007; Rual *et al.*, 2005)

Recently, samples have been collected whose members have both SNP genotyping data and expression data. These enable expression quantitative trait loci (eQTLs), that is, SNPs whose genotypes are significantly correlated with gene expression, to be detected (e.g., Dixon *et al.*, 2007; Monks *et al.*, 2004; Morley *et al.*, 2004). eQTLs can either be *cis* effects (if the associated SNP lies close to the gene with whose expression it is associated) or *trans* effects (the SNP can be associated with the expression of any gene in the genome). The definition of *cis* varies (e.g., <100 kb between SNP and gene in Dixon *et al.*, <2 Mb in Emilsson *et al.*, 2008), but in general, after correcting for multiple testing, considerably more significant *cis* effects are found than *trans* effects (Dixon *et al.*, 2007; Emilsson *et al.*, 2008; Schadt *et al.*, 2003).

There is growing evidence that eQTLs are associated with disease risk. For example, Emilsson *et al.* (2008) found that network modules derived from human and mouse adipose tissue expression data were enriched for eQTLs which were significantly associated with obesity-related traits in humans. A similar effect was found by Zhong *et al.* (2010), who initially identified $\sim$20,000 eSNPs significantly associated with gene expression in human liver, subcutaneous adipose, and omental adipose tissues. A gene-set enrichment analysis was then performed, using a modification of the method of Wang *et al.* (2007) on two independent type 2 diabetes GWAS datasets, using just the eSNPs showing *cis* effects. Several Kyoto encyclopedia of genes and genomes (KEGG) pathways were found to be significantly enriched in both studies. Pathways previously implicated in susceptibility to type 2 diabetes, such as PPAR signaling, were highlighted by this approach, and novel candidate pathways were also suggested. These results are promising, although it is unclear to what extent restricting the enrichment analysis to eSNPs improved its significance. Further evidence of the relationship between eQTLs and disease association was presented by Nicolae *et al.* (2010), who found that SNPs significantly associated with complex traits were more likely than expected by chance to also have significant eQTL *p*-values in lymphoblastoid cell lines on the HapMap samples (Duan *et al.*, 2008), with the excess being particularly pronounced for autoimmune disorders. Trait-associated SNPs were defined as those showing a *p*-value $<1 \times 10^{-5}$ in any published GWAS, and were taken from a publically available online database (http://www.genome.gov/gwastudies; Hindorff *et al.*, 2009). SNPs with the most significant eQTL *p*-values were found to be enriched for disease associations with Crohn's disease, Type I diabetes, and rheumatoid arthritis, but not type 2 diabetes, hypertension, coronary artery disease, or bipolar disorder, in the Wellcome Trust Case Control Consortium data.

There is considerable potential for using eQTL data to inform the results of GWAS, particularly with respect to discovering and interpreting biological pathways associated with disease risk, but several questions remain. For example, what are the relative utilities of *cis* and *trans* eQTLs? Furthermore, none of the studies relating eQTLs to GWAS results has used expression data from brain—such data may be particularly useful for brain-related diseases such as psychiatric disorders. For an excellent review of the relationships between gene expression networks and complex traits, see Schadt (2009).

It should be pointed out that networks based on molecular networks also contain potential pitfalls (Kraft and Raychaudhuri, 2009). For example, protein–protein interaction data are not always consistent, and correlation in gene expression may not necessarily capture important functional relationships. Care must be taken to ensure that the quality of the network data is sufficient to allow reliable conclusions to be drawn regarding disease involvement.

F. Text-mining methods

Like the network-based methods described above, text-mining methods seek to infer relationships between a list of interesting genes, SNPs, or genomic regions without relying on predefined pathways or ontologies. Text-mining methods work by looking for words (or "tokens") that appear more frequently in PubMed abstracts associated with genes on the list than in abstracts associated with other genes on the genotyping or expression array. The simplest way of testing whether a token is associated more frequently with genes on the list than with the remaining genes on the array is to assume that the number of genes on the list associated with the token follows a hypergeometric distribution. This assumes that each gene has the same a priori probability of being associated with a token (and that these probabilities are independent). This is not the case, as noted by Leong and Kipling (2009). First, some areas of biology have been more thoroughly studied than others, and genes in these areas would therefore be expected to appear on more abstracts, and thus more likely to be associated with tokens, than genes in less well-studied biological areas. Second, in general, the longer a gene has been known, the better-studied it tends to be. It is therefore important that text-mining methods account for such annotation bias. Leong and Kipling (2009) proposed three methods to analyze gene lists from differential expression studies that do this. They first calculated an empirical *p*-value for each token, based on creating 100,000 random gene lists matched for the number of genes and the number of associated abstracts. This was found to be accurate, but the large number of simulated gene lists required to obtain empirical *p*-values sufficiently small to withstand correction for the large number of tokens being tested made it computationally intensive. The second method tested for tokens with an abnormally high frequency of association with genes on the list given their frequency of association in the remainder of genes on the array. The third method adjusts the hypergeometric distribution to take into account the effect of increased annotation associated with the gene list on the probability of a gene on the list being associated with a particular token (and allowing this to vary between tokens). Both of the latter two methods were found to give similar results to simulation, but much faster, and are implemented on a public Web server PAKORA.

A text-mining method for analyzing GWAS data, called GRAIL, was developed by Raychaudhuri *et al.* (2009). This method takes a list of significantly associated SNPs and defines disease-associated *regions* to include all SNPs in LD ($r^2 > 0.5$) with the associated SNP, proceeding outwards in each direction to the nearest recombination hotspot (Myers *et al.*, 2005). For each gene in the genome, region, a vector of word counts is generated based on PubMed abstracts associated with that gene. For each gene in a region, a relatedness score is calculated with all other genes in the genome based on the similarity in their word counts, and these are ranked. The number of other regions containing a gene similar to the gene being tested is counted, and a *p*-value attached to that count. This is repeated for each gene in the region, and the significance of the region defined as the *p*-value of the best gene corrected by the number of genes in the region. The method thus not only defines closely related regions, but also which genes in the regions are themselves most closely related—these are the most likely candidates for biological involvement with disease susceptibility. Note that two genes do not have to appear in the same PubMed abstract in order to be given a high similarity score. The method is thus able to derive entirely novel networks of related genes. The authors applied the method to a set of validated SNPs associated with lipid levels, and found that it identified relationships more effectively than GO categories. They also applied it to a list of 74 SNPs with *p*-value $< 5 \times 10^{-5}$ from a GWA meta-analysis of Crohn's disease (Barrett *et al.*, 2008) and found that SNPs in regions implicated by their method were significantly more likely to be validated in a replication study than those in other regions. This is a convincing demonstration of the potential utility of pathway-based methods in identifying genes likely to be associated with disease. Since the method is based on regions, rather than SNPs or genes, it can also be used to analyze copy number variation (CNV) data, with the CNVs as the regions. An online version of GRAIL is available for public use.

GRAIL is an extremely promising method for pathway analysis of GWAS data. Since it analyzes regions defined by LD, it should be robust to LD between nearby genes (which would likely be in the same region). Relationships are only tested between genes in different regions, so the method should be robust to the positional clustering of functionally related genes noted by Hong *et al.* (2009). Annotation bias was corrected by only using PubMed abstracts published prior to December 2006 (before the recent wave of GWAS papers were published), but this may lose some useful information. Finally, it is not clear that GRAIL fully corrects for variable gene size, since large genes are more likely to have significant SNPs (and thus be included in the regions being tested). It is known that some biological areas are enriched for large genes (Hong *et al.*, 2009), so the possibility exists that this may be reflected in the relationships GRAIL detects—the extent to which this is true still needs to be assessed.

V. PUBLICALLY AVAILABLE PATHWAY DATABASES

Some of the more commonly used data sources for pathway analyses are described briefly in this section. For a more detailed review of available biological databases, see Stein (2003) and Tsui *et al.* (2007).

A. Gene ontology database

The GO database (Ashburner *et al.*, 2000; Harris *et al.*, 2004) is a collection of structured vocabularies, or ontologies, that describe and relate gene products in terms of their biological properties. There are three different ontologies: cellular components, biological processes and molecular functions, each containing a set of descriptors (GO terms), such as "steroid metabolism" or "vitamin D metabolic process." The cellular component ontology describes the locations of gene products within cells, for example, "nucleus," in terms of subcellular structures or macromolecular structures. The molecular function ontology describes activities that take place at the molecular level, such as "catalytic activity." Molecular function terms refer to the activities rather than the molecules that perform the actions, and do not specify where or when the action takes place. The biological process ontology contains terms that describe a series of events accomplished by one or more ordered assemblies of molecular functions, such as "signal transduction." It is sometimes difficult to distinguish between molecular functions and biological processes, but, in general, a biological process must contain more than one distinct step. Note that, technically, GO terms define gene sets, rather than pathways, since they do not represent dependencies between genes. GO is not a database of gene sequences, and it contains the properties of gene products, rather than the products themselves. A more detailed description of the ontologies, and their limitations, can be found on the GO Web site (http://www.geneontology.org/GO.doc.shtml).

An important feature of GO is that its terms are structured, that is, the set of genes corresponding to a more detailed GO term may be included within that corresponding to a more general GO term. Such relationships are denoted by the terms "is_a" and "part_of." These terms have different biological meanings, but the same meaning statistically: A is_a B and A part_of B both mean that if GO term A applies to a gene, so does GO term B. The annotation files relating genes to GO terms, which can be downloaded either from the GO Web site (http://www.geneontology.org/GO.current.annotations.shtml) or NCBI (ftp://ftp.ncbi.nih.gov/gene/DATA) do not include all these relationships. If these files are used as the basis of a pathway analysis, care must be taken to ensure that any missing gene-GO term relationships are added. This can be achieved by using the ontology files (http://www.geneontology.org/GO.downloads.ontology.shtml), which specify the relationships between the various GO terms (Holmans *et al.*, 2009). A graphical example (taken from the GO Web site) of the relationship structure between GO terms is shown in Fig. 7.2. The arrows represent "is_a" relationships, and connect "child" GO terms to "parent" GO terms. As can be seen, it is possible for a GO term to have more than one parent (e.g., GO:0042359, vitamin D metabolic process), and for a GO term to have more than one child (e.g., GO:0008152, metabolic process). As the arrows are followed up the graph, the GO terms become progressively more general, encompassing progressively larger sets of genes. For example, GO:0042359: vitamin D metabolic process contains 12 human gene products, whereas GO:0008202: steroid metabolic process contains 225 gene products and GO:0006629: lipid metabolic process contains 887 gene products. The sets of genes assigned to top-level GO terms may be very large indeed—GO:0008152: metabolic process contains 8396 gene products and GO:0008150: biological process contains no fewer than 14,934 gene products (as of April 22, 2010). Clearly, such large gene sets will not provide useful biological information. An advantage of the structure inherent in the GO ontologies is that researchers can select the GO terms they use in their analyzes according to the degree of specificity they require from their biological inferences, although it is not always obvious a priori what level of specificity will provide the most meaningful results. However, this structure also means that GO terms may be closely related to each other in terms of the genes they encompass. This can complicate the interpretation of results from pathway analyzes (Grossmann *et al.*, 2007).

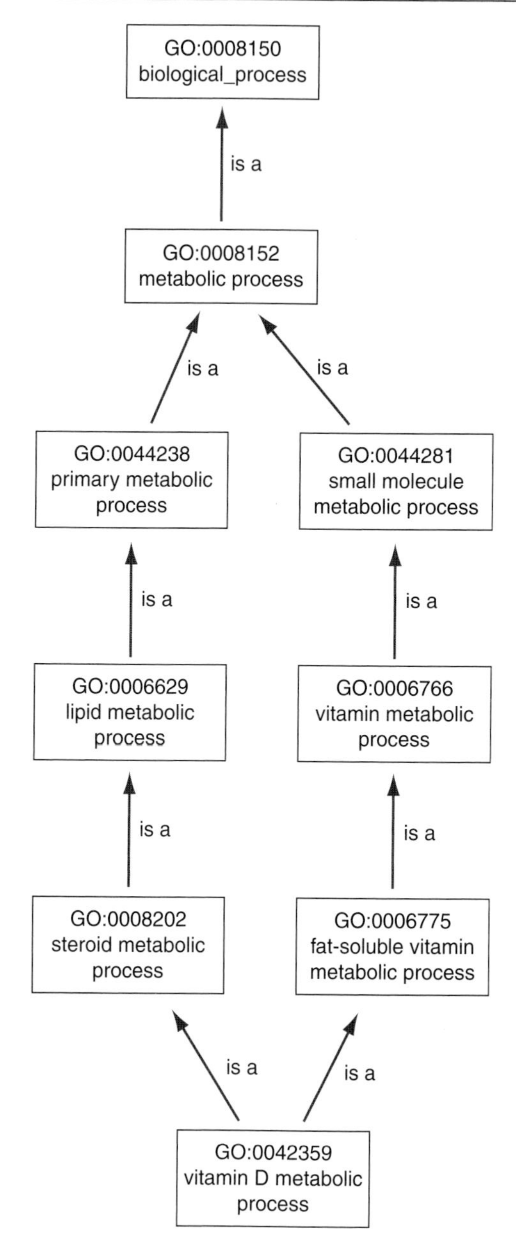

Figure 7.2. Example of the hierarchical structure of gene ontology (GO) terms. Arrows point from "child" terms to parent terms. The genes annotated to a "child" term are a subset of those annotated to a "parent" term.

One of the reasons the GO database is so commonly used as the basis of pathway analyses (see Khatri and Drăghici, 2005 for a review of analysis tools) is that it is very general and wide-ranging, with over 30,000 terms covering over 40 species. Inevitably, therefore, the quality of annotation will vary. Not all the assignments of GO terms to genes are based on experimental evidence—indeed, many are inferred solely from electronic annotations, without any expert human involvement. Of the 186,079 gene product—GO term annotations listed on the GO Web site (as of March 19th 2010), 97,707 (52.5%) were made electronically. Although the majority of these electronic annotations are reasonably accurate (Camon *et al.*, 2005), some are incorrect (King *et al.*, 2003). Of the remaining annotations, although all were manually checked, not all were based on experimental evidence—some are based on published *in silico* analysis (e.g., of the gene sequence). Indeed, as few as 1% of all gene-GO term annotations are based on experimental evidence, although this proportion varies considerably between species (Rhee *et al.*, 2008). One rather crude way of allowing for possibly unreliable annotations would be to exclude inferred electronic annotations entirely from pathway analyses based on GO. However, this would likely lose much useful information. A rather more satisfactory procedure would be to incorporate probabilistic estimates of gene function (e.g., Fraser and Marcotte 2004) into the analysis. However, more work is required before this is feasible.

B. Kyoto encyclopedia of genes and genomes

The KEGG pathway database currently (April 22, 2010) consists of 355 functional pathways defined across 129 eukaryotic genomes. Annotations linking genes to pathways can be downloaded separately for each organism (ftp://ftp.genome.jp/pub/kegg/genes/organisms/). Unlike GO, KEGG annotations are curated manually (Kanehisa and Goto, 2000), so they are likely to be more reliable. In addition, functional relationships within genes are shown (see Fig. 7.3 for an example), thus making it possible to test subsets of genes that are closely related. For these reasons, several pathway analyses use KEGG pathways (e.g., O'Dushlaine *et al.*, 2010; Wang *et al.*, 2007, 2009) rather than GO, despite the range of biology covered being smaller. In addition to the pathway database, there are 15 other databases (Kanehisa *et al.*, 2010) covering areas such as disease, drugs, and biochemical reactions. For more details on the KEGG database, visit the KEGG Web site (http://www.genome.jp/kegg/).

C. Panther

PANTHER (Protein Analysis THrough Evolutionary Relationships) is a database containing (as of April 22 2010) 165 canonical pathways capturing detailed biochemical interactions between proteins. Genes are also grouped into

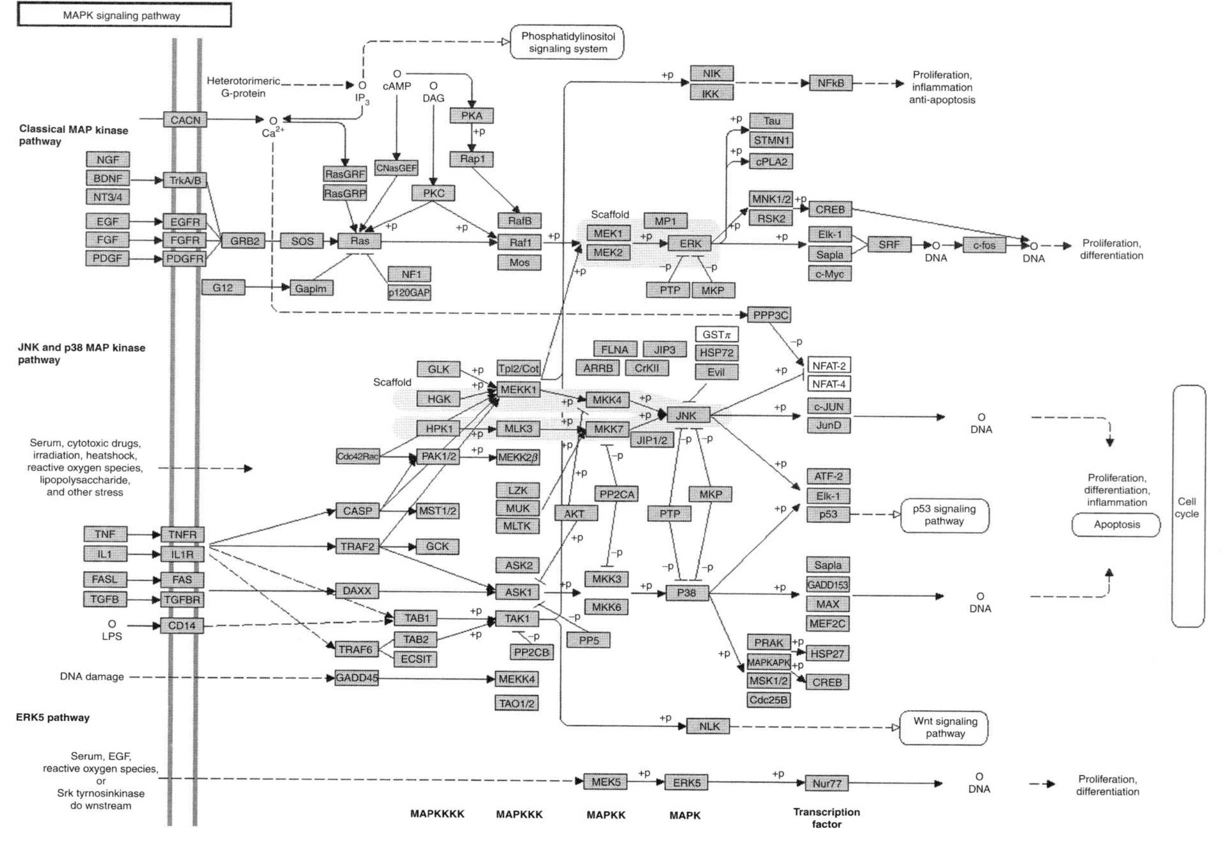

Figure 7.3. Example of a KEGG pathway. Here, the arrows indicate biological relationships.

molecular function and biological process ontologies, similar to GO. However, the number of ontology terms is greatly reduced compared to GO. Proteins are classified into families with similar function using genetic sequence data, based on 53 different species, experimental evidence where available and, uniquely, phylogenetic evolutionary information. The combination of these sources enables function to be predicted for proteins even in the absence of direct experimental data. These family groupings are then used to inform the construction of the pathways, which are manually curated. The protein families and pathways are available for download from the PANTHER Web site (http://www.pantherdb.org/) For more details, see Thomas *et al.* (2003) and Mi *et al.* (2010).

D. Molecular signatures database

The Molecular Signatures Database (MSIGDB; Subramanian *et al.*, 2005) is a collection of gene sets originally designed for use with the GSEA pathway analysis software. The most recent version (April 2008) contains 5452 gene sets, divided into five types. First, there are 386 positional gene sets, corresponding to each chromosome and each cytogenetic band containing at least one gene. Second, there are 1892 curated gene sets. Six hundred and thirty-nine of these are taken from pathway databases, and are usually curated by expert biologists. The remainder represents gene expression signatures of genetic and chemical perturbations, all cited in PubMed. Third, there are 837 motif gene sets, consisting of genes sharing a *cis*-regulatory motif that is conserved across the human, rat, mouse, and dog genomes (Xie *et al.*, 2005). Two hundred and twenty-two of these share a 3′-UTR microRNA binding motif, and 500 share a transcription factor binding site. Fourth, there are 883 computational gene sets, derived from cancer-oriented microarray data. Four hundred and fifty-six of these were taken from public databases by Segal *et al.* (2004), and found to be significantly dysregulated in cancer data. Four hundred and twenty-seven gene sets were defined by expression neighborhoods centered on 380 cancer-associated genes by Brentani *et al.* (2003). Fifth, there are 1454 GO gene sets (accession date January 25, 2008): 233 cellular component, 825 biological process and 396 molecular function. Only gene sets with 10 or more genes are included, and very large gene sets corresponding to very general GO terms are removed. GO terms are assigned to genes using experimental evidence only, and duplicated gene sets removed.

This is a very comprehensive collection of gene sets, covering a wide variety of areas of biology. As such, it is potentially very useful for performing pathway analyses. Although originally intended for use with the GSEA-P analysis packages, the gene sets are downloadable from the MSigDB Web site (http://www.broadinstitute.org/gsea/msigdb/index.jsp), and could thus be implemented in other pathway analyses. Potential drawbacks of the MSigDB gene sets are:

they may be of variable quality and more importantly, since the most recent version was compiled in April 2008, the gene sets compiled from public databases are likely to be out of date.

E. Database for annotation, visualization, and integrated discovery

Database for annotation, visualization, and integrated discovery (DAVID) contains an extremely comprehensive selection of annotations, including ontologies (GO and PANTHER), protein–protein interactions, pathways (including BioCarta, PANTHER, and KEGG), sequence features, expression data, and annotations from literature (PubMed). The unique feature of DAVID is its "knowledgebase," which integrates diverse gene identifiers and annotation categories to improve the annotation of genes. For example, use of the knowledgebase enabled 10–20% more GO terms to be assigned to genes, compared to annotations in each individual source. (Sherman *et al.*, 2007). DAVID will perform overrepresentation analysis on user-specified gene lists, based on the hypergeometric distribution, so is not designed for GWAS analysis. However, the knowledgebase is downloadable from the Web site (http://david.abcc.ncifcrf.gov/home.jsp) in a series of text files, enabling DAVID annotations to be used in other analysis programs.

F. The mouse genome informatics database

The mouse genome informatics (MGI) database (Bult *et al.*, 2008) is a comprehensive data source integrating genetic, genomic and phenotypic information about the laboratory mouse. It contains a number of databases relating genes disrupted in mouse experiments to a range of phenotypes (behavioral, molecular, and physiological), gene expression, and biochemical pathways. It also provides functional annotation on mouse genes using GO. Annotations are based on experimental data (either manually curated via the literature or directly contributed by researchers). There is also a resource for finding the orthologs of mouse genes in a variety of other mammals, including humans, thus making the gene sets and pathways more widely applicable. Gene sets corresponding to phenotypes and pathways can be downloaded from the Web site (http://www.informatics.jax.org/orthology.shtml).

This site provides information linking genes to phenotypes and biological processes that might not be directly testable in humans, and might thus be able to provide novel biological insights into human traits (Peters *et al.*, 2007) not available from the other data sources listed above. For example, Webber *et al.* (2009) applied the MGI phenotypic data to a set of CNVs associated with mental retardation in humans, and found a significant enrichment of genes whose mouse orthologs produced a phenotype related to the nervous system.

VI. UNRESOLVED ANALYTICAL ISSUES AND SCOPE FOR FURTHER RESEARCH

In this section, a number of currently unresolved analytical issues are discussed, and scope for further research indicated.

A. Which pathways/gene sets to use: Quality versus quantity

For a pathway analysis to be powerful, the set of genes in the pathway being tested needs to correspond fairly closely to the set of genes showing trait association. Genes or areas of biology that are poorly annotated are less likely to be highlighted by pathway analysis than those that are more comprehensively annotated. Clearly, more experimental work needs to be done to improve the overall biological annotation in the genome as a whole. In terms of statistical analysis, this argues for the use of large collections of pathways, such as those in GO, DAVID, or MSigDB, rather than smaller ones.

However, there are also some disadvantages associated with the use of large collections of pathways. First, a more stringent multiple testing correction is necessary if a large number of pathways are tested, making it harder to achieve the necessary significance, and thus reducing power to detect a pathway significantly enriched. Second, the amount of overlap in the sets of genes assigned to various pathways is likely to be greater when larger numbers of pathways are tested. This complicates both the multiple testing correction and the interpretation of the results, since if many closely related pathways are found to be enriched, it is not always clear which pathway is "driving" the results. This is a particular problem when using GO terms, since their hierarchical structure can result in GO terms being entirely contained within others. For these reasons, researchers often restrict the set of GO terms they test to a particular level in the GO hierarchy. For example, Wang *et al.* (2007) restricted analysis to terms at level 4 in the hierarchy (in Fig. 7.2, this would correspond to GO:0006629: lipid metabolic process and GO:0006766: vitamin metabolic process). Restrictions may also be made on the number of genes assigned to each GO term: Wang *et al.* restricted their analysis to terms assigned to between 10 and 200 genes, whereas the set of GO annotations in MSigDB is restricted to terms with at least 10 genes.

A rather more systematic method of restricting the set of GO terms is implemented by Alterovitz *et al.* (2007) in the Gene Ontology Partition Database (GO PaD). This selects a set of j GO terms covering the entire set of annotated genes (i.e., a partition of the ontology), such that the proportion of genes assigned to each node is as close to $1/j$ as possible—equivalent to maximizing the Shannon information (Shannon, 1948).

A problem with all such restrictions is that it is not clear a priori which level in the hierarchy, or the number of terms in the partition, best captures the biological information in the actual dataset being analyzed. It is possible to test several different hierarchy levels or partitions, but this would cause its own multiple testing problem. For this reason, some researchers prefer to test all GO terms, despite the stringent multiple testing correction (Holmans *et al.*, 2009). An interesting alternative way of reducing the set of pathways being tested is to "prune" pathways whose genes overlap by more than a fixed threshold with a more significantly enriched pathway. The measure of overlap could be defined as the number of genes common to both pathways divided by the number of genes in either pathway. Such "intelligent pruning" would ensure that the set of pathways retained in the analysis would be the ones that best capture the association signal in the data. However, this approach may cause problems for performing multiple testing correction. The optimal way of reducing the set of pathways to be tested while still capturing the underlying biology of the dataset is an area that requires further work.

Even when the set of pathways has been decided, relationships between them may complicate the interpretation of enrichment results. In particular, the parent–child relationships in the GO ontologies may cause false-positive results unrelated to disease biology (Grossmann *et al.*, 2007). To address this problem, Grossmann *et al.* developed a method for testing whether a GO term is overrepresented on a gene list given the number of genes its parent terms contribute to the list. This approach, implemented in the package Ontologizer (http://compbio.charite.de/index.php/ontologizer2.html) was found to give fewer false positives than the standard term-by-term analysis. However, it is designed to analyze gene lists (e.g., from microarray data) and thus needs further development for application to GWAS data.

When selecting pathways, researchers should also bear in mind that large pathways tend to have a greater chance of being statistically significant when GWAS data are analyzed using analysis tools designed for microarray data (Elbers *et al.*, 2009). Furthermore, Elbers *et al.* also noted that the same pathways analyzed with different analysis tools often gave very different results. Sometimes, this is due to the pathways being different, despite having similar names. However, even pathways that should be the same (e.g., KEGG pathways) occasionally gave different results. Thus, consistent annotation across databases is vitally important, as is ensuring that the databases are kept up to date. As was noted earlier, the quality of annotation varies greatly between pathways (e.g., some annotations of GO terms to genes are validated experimentally, others are inferred electronically). Ideally, pathway analyses should take such differences into account.

B. Correcting for multiple testing

Standard methods for correcting multiple hypothesis tests, such as Bonferroni or Sidak are conservative, since they do not allow for the dependence between pathways, which can be quite large, especially when using GO. False-discovery rate (FDR) methods (e.g., Benjamini and Hochberg, 1995), which control the proportion of events reported as significant that are actually false positives, are a more appropriate method for multiple testing correction, and are implemented in many pathway analysis tools. However, even these make the assumption that the significance tests for each pathway are independent. If the method used to select the set of pathways being tested is related to the enrichment test, then none of these methods is applicable. For example, Holmans *et al.* (2009) restricted the set of GO terms being tested to those with at least two genes on the list of significant genes (in order to prevent small pathways being called significantly enriched on the basis of one, possibly chance, significant SNP). Clearly, this will shift the expected distribution of reported *p*-values toward significance, and any multiple testing correction applied must reflect this shift. Holmans *et al.* (2009) therefore used a bootstrapping procedure, as described earlier, in which one of the random gene lists was selected as the "real data" and tested against a sample drawn with replacement from the remaining gene lists, with the same selection of GO terms being applied. This procedure also gives a significance test for the number of enriched GO terms, although this will be influenced by which genes are significant in the real dataset. If these have been well-studied, and annotated to several closely related GO terms, the significance of the number of enriched terms will be inflated. This bias may be reduced by "pruning" out closely related pathways, but the performance of the approach needs to be evaluated further.

C. Assigning SNPs to genes

This issue applies to analyses of GWAS data only. Many researchers assign SNPs to genes not only if they physically lie within the boundaries of the gene, but also if they lie within a certain distance either side, often called a "gene window." The idea of allowing a window is to capture proximal functional elements (such as regulatory regions), which often lie close to, but not within, genes. The optimal size of window is an open question. Holmans *et al.* (2009) used a 20 kb window, motivated by a study (Veyrieras *et al.*, 2008), which showed that most eQTLs lie within 20 kb of genes. However, other window sizes have been used, for example, 5 kb (Torkamani *et al.*, 2008) or 500 kb (Wang *et al.*, 2007). The larger the window, the more genes will have SNPs assigned to them, and thus contribute to the analysis, and the more functional elements will be covered. However, large windows also increase the chance that a SNP has no functional relationship with the gene to which it is assigned. Large windows also increase the overlap between genes, increasing the number of SNPs that could be assigned to more than one gene. Some studies have assigned such SNPs to the nearest gene (e.g., Torkamani *et al.*, 2008), while others have discarded them completely (e.g., Askland *et al.*, 2009). For studies where SNPs are assigned to all genes within whose windows they lie (e.g., Holmans *et al.*, 2009), it is important that the statistical analysis method allows for window overlap. Methods based on permuting the phenotypes and reanalyzing the GWAS will do this automatically.

Recently, methods have been devised to increase the coverage of genotyping chips by imputing genotypes at SNPs not on the chip probabilistically (e.g., Marchini *et al.*, 2007). This will greatly increase the coverage of genes, particularly small genes (Hong *et al.*, 2009), and should therefore increase the power of pathway analyses (Holmans *et al.*, 2009). There is some evidence that the increased gene coverage from imputation removes some of the benefits of using a gene window (Holmans *et al.*, 2009), but this needs further study. If HapMap data (International HapMap Consortium, 2003) is used for imputation, the dataset will typically consist of 1–2 million SNPs. If the new 1000 Genomes data are used, more than 5 million SNPs will become available. This will vastly increase the computational burden on methods which rely on permutation and reanalysis of the genome.

D. Correcting for dependence between and within genes

For studies where there is only one measurement per gene (e.g., gene expression studies), correction for dependence between observations within genes is not necessary. However, observations from different genes may be correlated, and methods for testing enriched pathways should allow for this correlation. A natural way to do this is to randomly permute phenotypes between individuals and repeat the expression analysis. However, of the commonly used packages, only GSEA (Subramanian *et al.*, 2005) and SAFE (Barry *et al.*, 2005) appear to use permutation methods to obtain significance.

For GWAS, there are multiple observations (SNPs) per gene, which makes corrections within genes for multiple observations necessary. Furthermore, these observations are correlated due to LD, making standard methods for multiple testing correction (e.g., Bonferroni) unsuitable. Methods that employ phenotype permutation and reanalysis of the whole genome (e.g., O'Dushlaine *et al.*, 2010; Wang *et al.*, 2009) will account for these correlations automatically. However, these are computationally intensive and require the availability of individual genotyping data. A number of methods have been proposed for application to summary data (e.g., association *p*-values): Holmans *et al.* (2009) use only the most significant SNP per gene, which removes the need to consider multiple correlated observations per gene. Correction for multiple SNPs per gene is performed by generating random gene lists sampling SNPs (rather than genes) at random, reflecting the fact that large genes with many SNPs are more likely to have significant SNPs by chance. This approach implicitly assumes that the relationship between the number of SNPs in a gene and the effective number of independent tests, which depends on the LD between SNPs, is approximately constant between pathways, and will be conservative for pathways whose genes contain several SNPs in tight LD. Hong *et al.* (2009) cluster SNPs if $r^2 > 0.8$, and then apply a Bonferroni correction for the number of clusters. This is an appealing approach. However, the high criterion on r^2 for forming clusters means that there could be considerable LD between clusters, making the Bonferroni approximation conservative. Relaxing the r^2 criterion would increase the number of independent signals within clusters, making the Bonferroni correction anticonservative. Essentially, the Hong *et al.* method is making the same assumption as Holmans *et al.*, that the degree of LD between SNPs is approximately constant between pathways.

There is also the issue of correcting for LD between genes. Again, permuting the phenotype and reanalyzing the genome will correct for this automatically. Otherwise, Holmans *et al.* (2009) suggest removing regions exhibiting long-range LD (e.g., the MHC region) and repeating the pathway analysis. This will probably be conservative. Hong *et al.* (2009) drop genes in LD with more significant genes. Again, this could reduce power if the gene that is dropped is not in the same pathway as the gene that is retained. Neither of these methods is ideal. Perhaps the best practical method for correcting for inter-gene LD without resorting to whole-genome permutation is to manually inspect which genes are contributing to the signal in significantly enriched pathways, collapse together genes that are close together (or in high LD) in the same pathway, and repeat the analysis. This raises the question of what criteria on distance or LD to use for the collapsing. Clearly, there is considerable work needed to develop and test methods for correcting for LD between and within genes in the absence of whole-genome permutation.

Even if whole-genome permutation is used, it is advisable to check that the enrichment signal in a pathway is not coming from one gene (O'Dushlaine *et al.*, 2010). This is a particular problem for methods that use the association *p*-values, such as GSEA or self-contained tests such as those based on the product of *p*-values, which can be dominated by a single gene (Hong *et al.*, 2009). For example, the association of late-onset Alzheimer's disease to APOE is extremely strong, so one must be careful that any significant pathway enrichment is not being driven solely by signal from the APOE region.

E. What gene/pathway-wide association measure to use?

Many analysis methods performing *competitive* pathway enrichment tests use the most significant *p*-value as the gene-wide measure of association (if there is only one observation per gene, this is the only measure available). Note that the corrections for multiple observations per genes, as described above, may mean that the list of *N* significant genes (for overrepresentation analyses) does not necessarily comprise the genes with the *N* most significant single-SNP *p*-values, or that the genes are not ranked (for gene-set enrichment analyses) in order of their most significant SNPs. An exception is the method of Holmans *et al.* (2009). The SNP-ratio test of O'Dushlaine *et al.* (2010) uses the proportion of nominally significant ($p < 0.05$) SNPs as its gene-wide statistic, thereby counting multiple independent signals in genes.

For *self-contained* tests, there are a wide variety of gene- and pathway-wide measures, such as Hotelling's T^2, Fisher's method and truncated product measures—these were reviewed in a previous section. All of these require phenotype permutation and whole-genome reanalysis to obtain *p*-values, when applied to GWAS data, in order to allow for the correlation between SNPs. Strictly speaking, permutations should also be used to obtain significance when these methods are applied to gene expression data.

The power of GSEA was compared with various self-contained tests on expression data by Ackermann and Strimmer (2009), and generally found to be inferior. However, since self-contained tests test the null hypothesis that there is no association between the pathway and the phenotype, they are not robust to systematic genome-wide inflation of single-gene or single-SNP test statistics, which are particularly common in GWAS (Devlin and Roeder, 1999). Such inflation does not affect competitive tests, provided it is not pathway dependent. The best way to correct self-contained tests for systematic inflation of test statistics is still unclear. The power of overrepresentation analyses based on lists of significant genes compared to methods based on averaging test statistics across genes was investigated by Newton *et al.* (2007). They found that overrepresentation analysis was more powerful when the excess proportion of true differentially expressed genes in the pathway over that in the dataset as a whole was relatively small, but the effects of these genes were large. Conversely, averaging the test statistics was more powerful when the pathway contained a large proportion of differentially expressed genes of small effect. No power comparison of pathway methods has yet been performed for GWAS data, but the relative performance of the methods is likely to depend on the genetic architecture of the trait.

F. How long should the list of significant genes be?

For overrepresentation analysis, a list of significant genes must be defined. Clearly, the threshold used to define this list will be of key importance in determining the power of the study. If the gene list is too short, there are unlikely to be enough genes in any pathway for power to detect enrichment to be high. Conversely, if the gene list is too long, the proportion of significant genes in most pathways will be high, reducing the power of pathway analyses to discriminate between them. The optimal length of gene list is likely to depend on the proportion of truly trait-associated genes, which is unknown. In practice, studies have used differing definitions of gene list: for example, Hong *et al.* (2009) used the top 1% of genes, while Torkamani *et al.* (2008) used the top 2.5%. Baranzini *et al.* (2009) used genes containing a SNP with $p < 0.05$, although they restricted their gene list to genes on a protein interaction network. Holmans *et al.* (2009) used a number of different SNP-wise p-value cutoffs to define significant genes, selecting the one giving the most significant excess of enriched pathways. It is clear that more work needs to be done to inform the choice of gene lists for overrepresentation analyses.

G. Are whole-genome permutations necessary?

Permuting phenotype among individuals and repeating the whole genome analysis is desirable from a statistical point of view, since it corrects for LD within and between genes. However, this is computationally intensive, and requires access to the individual genotyping data. When performing the permutations, it is necessary to respect the structure of the data. For example, one should try and carry out permutations within strata assumed to be homogeneous with respect to ancestry (Kraft and Raychaudhuri, 2009). This requires adjustment for covariates, such as those for population stratification (Price *et al.*, 2006), if this was done in the original study, further increasing computing time. There is evidence that imputing SNPs improves the power of pathway analyses (Holmans *et al.*, 2009), but this will increase the computing burden still further, by greatly increasing the number of SNPs to be analyzed. The power of pathway analysis is greatly increased by analyzing large samples (Holmans *et al.*, 2009), such as those of the International Schizophrenia Consortium and, most recently, the Psychiatric Genetics Consortium (Psychiatric GWAS Consortium Steering Committee, 2009).

For these reasons, it is often impossible to perform enough whole genome permutation analyses to get empirical estimates of the p-values for pathway enrichment sufficiently small to withstand multiple testing correction for the number of pathways being tested, particularly if large sets of pathways are being tested (e.g., gene ontology, or MSigDB). Some authors have attempted to resolve this problem by replicating the enriched pathways across multiple independent datasets (O'Dushlaine *et al.*, 2010; Wang *et al.*, 2009). However, it is unclear how the power of this approach compares to performing a single analysis on the combined samples. In the context of genome-wide association analysis based on single SNPs, Skol *et al.* (2006) found that the latter approach was considerably more powerful than the former. However, more research is needed to ascertain whether a similar conclusion holds for pathway analyses. It is also important that more work is done to compare the statistical performance of methods that do not require whole-genome permutations to those that do, in order to see whether the former are sufficiently statistically accurate to yield useful results.

VII. SOFTWARE PACKAGES FOR PATHWAY ANALYSIS OF GWAS DATA

Several software packages exist for performing pathway analysis on gene expression data, reviewed in detail by Huang da *et al.* (2009b). A list of publically available packages for analyzing genome-wide association data is given in Table 7.1. For each package, the analysis method it employs (GSEA, overrepresentation, etc) is given, together with a Web site for download.

VIII. CONCLUSIONS

Pathway analyses are well-established in the analysis of expression data, yielding important insights into the underlying biology of disease. Recently, pathway analysis methods have been developed for the analysis of genome-wide association data. While initial results of such analyses are promising, there remain several statistical issues to address. When performing a pathway analysis of genome-wide association data, it is vital that the analysis method corrects for variable numbers of SNPs per gene, allowing for LD between these SNPs. The best methods for doing this accurately are as yet unclear. Permuting the trait phenotype among individuals and reanalyzing the whole genome is statistically correct, but requires considerable computational resources, and also access to the individual genotype data. Therefore, the performance of other methods is currently under study.

The most appropriate pathway analysis method will depend on whether the researcher wishes to carry out a competitive test (testing whether enrichment for association signal is greater in the pathway than in the genome as a whole) or a self-contained test (testing whether there is any enrichment of association signal in the pathway). Methods such as overrepresentation analysis or GSEA address the former, while multivariate methods such as Fisher's method for combining p-values address the latter. The relative power of these methods has yet to be determined. It is important to note that self-contained tests are affected by systemic genome-wide inflation of the association test statistics (e.g., due to population stratification), and the optimal way of correcting for this is still unclear.

The choice of pathways to analyze is important—a large collection, such as GO, increases the chance of a pathway coinciding with the disease susceptibility genes in the actual data, but the quality of annotation in such collections can be variable, since it is not always based on experimental data. Currently, there is no way to allow for variable annotation methods in pathway analyses, short of removing lower-quality annotations entirely. Annotations are continually changing as new experimental data become available. Therefore, whatever pathways are chosen, it is important that annotations are up to date.

Table 7.1. Programs for Pathway Analysis of Genome-Wide Association Data

Program	References	Web site	Analysis method	Gene-wide association measure	Individual genotype data needed?
ALIGATOR	Holmans *et al.* (2009)	http://x004.psycm.uwcm.ac.uk/~peter/	Overrepresentation analysis	Best single-SNP *p*-value	No
GenGen	Wang *et al.* (2009)	http://www.openbioinformatics.org/gengen/	Gene-set enrichment	Best single-SNP *p*-value	Yes
ProxyGeneLD	Hong *et al.* (2009)	http://ki.se/ki/jsp/polopoly.jsp?d=26072&a=67192&l=en	None[a]	Best single-SNP *p*-value	No
GSEA-SNP	Holden *et al.* (2008)	http://www.nr.no/pages/samba/area_emr_smbi_gseasnp	Gene-set enrichment[b]	N/A	Yes
SRT	O'Dushlaine *et al.* (2010)	https://sourceforge.net/projects/snpratiotest/	Overrepresentation analysis	Proportion of SNPs with $p < 0.05$	Yes
EVA	Askland *et al.* (2009)	http://www.exploratoryvisualanalysis.org/	Overrepresentation analysis	User—defined[c]	No
GRAIL	Raychaudhuri *et al.* (2009)	http://www.broad.mit.edu/mpg/grail	Text-based analysis	Region-wide best single-SNP *p*-value	No

[a]Produces a gene list for input into other analysis packages.
[b]Based on SNPs rather than genes.
[c]User needs to apply correction for multiple SNPs per gene before inputting list.

In addition to using predefined pathways from databases, such as GO and KEGG, there is considerable interest in integrating other data sources, in order to obtain novel biological insights. Text-mining is a promising way of achieving this aim, as is the mining of protein interaction networks. Recently, evidence has been presented that SNPs which influence gene expression levels (eSNPs) may have an important role in disease susceptibility. eSNPs could be used to define new pathways, or to weight association evidence in existing pathways. There is considerable scope for work in this area.

To conclude, pathway analyses are an extremely promising way of extracting biological meaning from genome-wide data. However, there are several as yet unsolved statistical issues, particularly in the analysis of genome-wide association data, and it is likely that this will be a productive area of research over the next few years.

References

Ackermann, M., and Strimmer, K. (2009). A general modular framework for gene set enrichment analysis. *BMC Bioinform.* **10,** 47.

Allison, D. B., Cui, X., Page, G. P., and Sabripour, M. (2006). Microarray data analysis: From disarray to consolidation and consensus. *Nat. Rev. Genet.* **7,** 55–65.

Alterovitz, G., Xiang, M., Mohan, M., and Ramoni, M. F. (2007). GO PaD: The gene ontology partition database. *Nucleic Acids Res.* **35**(Database issue), D322–D327.

Ashburner, M., Ball, C. A., Blake, J. A., Botstein, D., Butler, H., Cherry, J. M., Davis, A. P., Dolinski, K., Dwight, S. S., Eppig, J. T., *et al.* (2000). Gene ontology: Tool for the unification of biology. The Gene Ontology Consortium. *Nat. Genet.* **25,** 25–29.

Askland, K., Read, C., and Moore, J. (2009). Pathways-based analyses of whole-genome association study data in bipolar disorder reveal genes mediating ion channel activity and synaptic neuro-transmission. *Hum. Genet.* **125,** 63–79.

Baranzini, S. E., Galwey, N. W., Wang, J., Khankhanian, P., Lindberg, R., Pelletier, D., Wu, W., Uitdehaag, B. M., and Kappos, L. GeneMSA Consortium (2009). Pathway and network-based analysis of genome-wide association studies in multiple sclerosis. *Hum. Mol. Genet.* **18,** 2078–2090.

Barrett, J. C., Hansoul, S., Nicolae, D. L., Cho, J. H., Duerr, R. H., Rioux, J. D., Brant, S. R., Silverberg, M. S., Taylor, K. D., Barmada, M. M., *et al.* (2008). Genome-wide association defines more than 30 distinct susceptibility loci for Crohn's disease. *Nat. Genet.* **40,** 955–962.

Barry, W. T., Nobel, A. B., and Wright, F. A. (2005). Significance analysis of functional categories in gene expression studies: A structured permutation approach. *Bioinformatics* **21,** 1943–1949.

Beissbarth, T., and Speed, T. P. (2004). GOstat: Find statistically overrepresented gene ontologies within a group of genes. *Bioinformatics* **20,** 1464–1465.

Benjamini, Y., and Hochberg, Y. (1995). Controlling the false discovery rate: A practical an powerful approach to multiple testing. *J. R. Stat. Soc. Series B* **57,** 289–300.

Breitling, L. P., Steyerberg, E. W., and Brenner, H. (2009). The novel "genomic pathway approach" to complex diseases: A reason for (over-)optimism? *Epidemiology* **20,** 500–507.

Brentani, H., Caballero, O. L., Camargo, A. A., da Silva, A. M., da Silva, W. A., Jr, DiasNeto, E., Grivet, M., Gruber, A., Guimaraes, P. E., Hide, W., *et al.* (2003). The generation and utilization of a cancer-oriented representation of the human transcriptome by using expressed sequence tags. *Proc. Natl. Acad. Sci. USA* **100,** 13418–13423.

Bugrim, A., Nikolskaya, T., and Nikolsky, Y. (2004). Early prediction of drug metabolism and toxicity: Systems biology approach and modeling. *Drug Discov. Today* **9**, 127–135.
Bult, C. J., Eppig, J. T., Kadin, J. A., Richardson, J. E., and Blake, J. A. Mouse Genome Database Group (2008). The Mouse Genome Database (MGD): Mouse biology and model systems. *Nucleic Acids Res.* **36**(Database issue), D724–D728.
Camon, E. B., Barrell, D. G., Dimmer, E. C., Lee, V., Magrane, M., Maslen, J., Binns, D., and Apweiler, R. (2005). An evaluation of GO annotation retrieval for BioCreAtIvE and GOA. *BMC Bioinform.* **6**(Suppl. 1), S17.
Cantor, R. M., Lange, K., and Sinsheimer, J. S. (2010). Prioritizing GWAS results: A review of statistical methods and recommendations for their application. *Am. J. Hum. Genet.* **86,** 6–22.
Conti, D. V., Cortessis, V., Molitor, J., and Thomas, D. C. (2003). Bayesian modeling of complex metabolic pathways. *Hum. Hered.* **56,** 83–93.
De la Cruz, O., Wen, X., Ke, B., Song, M., and Nicolae, D. L. (2010). Gene, region and pathway level analyses in whole-genome studies. *Genet. Epidemiol.* **34,** 222–231.
Dennis, G., Jr, Sherman, B. T., Hosack, D. A., Yang, J., Gao, W., Lane, H. C., and Lempicki, R. A. (2003). DAVID: Database for annotation, visualization, and integrated discovery. *Genome Biol.* **4,** P3.
Devlin, B., and Roeder, K. (1999). Genomic control for association studies. *Biometrics* **55,** 997–1004.
Dixon, A. L., Liang, L., Moffatt, M. F., Chen, W., Heath, S., Wong, K. C., Taylor, J., Burnett, E., Gut, I., Farrall, M., *et al.* (2007). A genome-wide association study of global gene expression. *Nat. Genet.* **39,** 1202–1207.
Dobrin, R., Zhu, J., Molony, C., Argman, C., Parrish, M. L., Carlson, S., Allan, M. F., Pomp, D., and Schadt, E. E. (2009). Multi-tissue coexpression networks reveal unexpected subnetworks associated with disease. *Genome Biol.* **10,** R55.
Duan, S., Huang, R. S., Zhang, W., Bleibel, W. K., Roe, C. A., Clark, T. A., Chen, T. X., Schweitzer, A. C., Blume, J. E., Cox, N. J., *et al.* (2008). Genetic architecture of transcript-level variation in humans. *Am. J. Hum. Genet.* **82,** 1101–1113.
Dudbridge, F., and Koeleman, B. P. (2003). Rank truncated product of *P*-values, with application to genomewide association scans. *Genet. Epidemiol.* **25,** 360–366.
Dudbridge, F., and Koeleman, B. P. (2004). Efficient computation of significance levels for multiple associations in large studies of correlated data, including genomewide association studies. *Am. J. Hum. Genet.* **75,** 424–435.
Elbers, C. C., van Eijk, K. R., Franke, L., Mulder, F., van der Schouw, Y. T., Wijmenga, C., and Onland-Moret, N. C. (2009). Using genome-wide pathway analysis to unravel the etiology of complex diseases. *Genet. Epidemiol.* **33,** 419–431.
Emilsson, V., Thorleifsson, G., Zhang, B., Leonardson, A. S., Zink, F., Zhu, J., Carlson, S., Helgason, A., Walters, G. B., Gunnarsdottir, S., *et al.* (2008). Genetics of gene expression and its effect on disease. *Nature* **452,** 423–428.
Ferreira, M. A., O'Donovan, M. C., Meng, Y. A., Jones, I. R., Ruderfer, D. M., Jones, L., Fan, J., Kirov, G., Perlis, R. H., Green, E. K., *et al.* (2008). Collaborative genome-wide association analysis supports a role for ANK3 and CACNA1C in bipolar disorder. *Nat. Genet.* **40,** 1056–1058.
Fisher, R. A. (1932). Statistical methods for research workers. Oliver and Boyd, London.
Fraser, A. G., Marcotte, E. M. (2004). A probabilistic view of gene function. *Nat. Genet.* **36,** 559–564
Glazko, G. V., and Emmert-Streib, F. (2009). Unite and conquer: Univariate and multivariate approaches for finding differentially expressed gene sets. *Bioinformatics* **25,** 2348–2354.
Goeman, J. J., and Bühlmann, P. (2007). Analyzing gene expression data in terms of gene sets: Methodological issues. *Bioinformatics* **23,** 980–987.
Goeman, J. J., van de Geer, S. A., de Kort, F., and van Houwelingen, H. C. (2004). A global test for groups of genes: Testing association with a clinical outcome. *Bioinformatics* **20,** 93–99.

Grossmann, S., Bauer, S., Robinson, P. N., and Vingron, M. (2007). Improved detection of overrepresentation of Gene-Ontology annotations with parent child analysis. *Bioinformatics* **23,** 3024–3031.

Harris, M. A., Clark, J., Ireland, A., Lomax, J., Ashburner, M., Foulger, R., Eilbeck, K., Lewis, S., Marshall, B., Mungall, C., *et al.* (2004). The Gene Ontology (GO) database and informatics resource. *Nucleic Acids Res.* **32**(Database issue), D258–D261.

Hindorff, L. A., Sethupathy, P., Junkins, H. A., Ramos, E. M., Mehta, J. P., Collins, F. S., and Manolio, T. A. (2009). Potential etiologic and functional implications of genome-wide association loci for human diseases and traits. *Proc. Natl. Acad. Sci. USA* **106,** 9362–9367.

Holden, M., Deng, S., Wojnowski, L., and Kulle, B. (2008). GSEA-SNP: Applying gene set enrichment analysis to SNP data from genome-wide association studies. *Bioinformatics* **24,** 2784–2785.

Holmans, P., Green, E. K., Pahwa, J. S., Ferreira, M. A., Purcell, S. M., Sklar, P., Wellcome Trust Case Control Consortium, Owen, M. J., O'Donovan, M. C., and Craddock, N. (2009). Gene ontology analysis of GWA study data sets provides insights into the biology of bipolar disorder. *Am. J. Hum. Genet.* **85,** 13–24.

Hong, M. G., Pawitan, Y., Magnusson, P. K., and Prince, J. A. (2009). Strategies and issues in the detection of pathway enrichment in genome-wide association studies. *Hum. Genet.* **126,** 289–301.

Hosack, D. A., Dennis, G., Jr., Sherman, B. T., Lane, H. C., and Lempicki, R. A. (2003). Identifying biological themes within lists of genes with EASE. *Genome Biol.* **4,** R70.

Huang da, W., Sherman, B. T., and Lempicki, R. A. (2009a). Systematic and integrative analysis of large gene lists using DAVID bioinformatics resources. *Nat. Protoc.* **4,** 44–57.

Huang da, W., Sherman, B. T., and Lempicki, R. A. (2009b). Bioinformatics enrichment tools: Paths toward the comprehensive functional analysis of large gene lists. *Nucleic Acids Res.* **37,** 1–13.

Hummel, M., Meister, R., and Mansmann, U. (2008). GlobalANCOVA: Exploration and assessment of gene group effects. *Bioinformatics* **24,** 78–85.

Ideker, T., Ozier, O., Schwikowski, B., and Siegel, A. F. (2002). *Bioinformatics* **18**(Suppl. 1), S233–S240.

International HapMap Consortium (2003). The International HapMap Project. *Nature* **426,** 789–796.

International Schizophrenia Consortium Purcell, S. M., Wray, N. R., Stone, J. L., Visscher, P. M., O'Donovan, M. C., Sullivan, P. F., and Sklar, P. (2009). Common polygenic variation contributes to risk of schizophrenia and bipolar disorder. *Nature* **460,** 748–752.

Jiang, Z., and Gentleman, R. (2007). Extensions to gene set enrichment. *Bioinformatics* **23,** 306–313.

Kanehisa, M., and Goto, S. (2000). KEGG: Kyoto encyclopedia of genes and genomes. *Nucleic Acids Res.* **28**(1), 27–30.

Kanehisa, M., Goto, S., Furumichi, M., Tanabe, M., and Hirakawa, M. (2010). KEGG for representation and analysis of molecular networks involving diseases and drugs. *Nucleic Acids Res.* **38** (Database issue), D355–D360.

Khatri, P., and Drăghici, S. (2005). Ontological analysis of gene expression data: Current tools, limitations, and open problems. *Bioinformatics* **21,** 3587–3595.

King, O. D., Foulger, R. E., Dwight, S. S., and White, J. V. (2003). Roth FP (2003) Predicting gene function from patterns of annotation. *Genome Res.* **13**(5), 896–904.

Kooperberg, C., and Ruczinski, I. (2005). Identifying interacting SNPs using Monte Carlo logic regression. *Genet. Epidemiol.* **28,** 157–170.

Kraft, P., and Raychaudhuri, S. (2009). Complex diseases, complex genes. Keeping pathways on the right track. *Epidemiology* **20,** 508–510.

Lage, K., Karlberg, E. O., Størling, Z. M., Olason, P. I., Pedersen, A. G., Rigina, O., Hinsby, A. M., Tümer, Z., Pociot, F., Tommerup, N., *et al.* (2007). A human phenome-interactome network of protein complexes implicated in genetic disorders. *Nat. Biotechnol.* **25,** 309–316.

Lein, E. S., Hawrylycz, M. J., Ao, N., Ayres, M., Bensinger, A., Bernard, A., Boe, A. F., Boguski, M. S., Brockway, K. S., Byrnes, E. J., *et al.* (2007). Genome-wide atlas of gene expression in the adult mouse brain. *Nature* **445,** 168–176.

Leong, H. S., and Kipling, D. (2009). Text-based over-representation analysis of microarray gene lists with annotation bias. *Nucleic Acids Res.* **37,** e79.

Lesnick, T. G., Papapetropoulos, S., Mash, D. C., Ffrench-Mullen, J., Shehadeh, L., de Andrade, M., Henley, J. R., Rocca, W. A., Ahlskog, J. E., and Maraganore, D. M. (2007). A genomic pathway approach to a complex disease: Axon guidance and Parkinson disease. *PLoS Genet.* **3,** e98.

Lu, Y., Liu, P. Y., Xiao, P., and Deng, H. W. (2005). Hotelling's T2 multivariate profiling for detecting differential expression in microarrays. *Bioinformatics* **21,** 3105–3113.

Maher, B. (2008). Personal genomes: The case of the missing heritability. *Nature* **456,** 18–21.

Makambi, K. H. (2003). Weighted inverse chi-square method for correlated significance tests. *J. Appl. Stat.* **30,** 225–234.

Manolio, T. A., Brooks, L. D., and Collins, F. S. (2008). A HapMap harvest of insights into the genetics of common disease. *J. Clin. Invest.* **118,** 1590–1605.

Marchini, J., Howie, B., Myers, S., McVean, G., and Donnelly, P. (2007). A new multipoint method for genome-wide association studies by imputation of genotypes. *Nat. Genet.* **39**(7), 906–913.

Mi, H., Dong, Q., Muruganujan, A., Gaudet, P., Lewis, S., Thomas, P. D. (2010). PANTHER version 7: improved phylogenetic trees, orthologs and collaboration with the Gene Ontology Consortium. *Nucl. Acids. Res.* **38,** D204–D210.

Monks, S. A., Leonardson, A., Zhu, H., Cundiff, P., Pietrusiak, P., Edwards, S., Phillips, J. W., Sachs, A., and Schadt, E. E. (2004). Genetic inheritance of gene expression in human cell lines. *Am. J. Hum. Genet.* **75,** 1094–1105.

Mootha, V. K., Lindgren, C. M., Eriksson, K. F., Subramanian, A., Sihag, S., Lehar, J., Puigserver, P., Carlsson, E., Ridderstråle, M., Laurila, E., *et al.* (2003). PGC-1alpha-responsive genes involved in oxidative phosphorylation are coordinately downregulated in human diabetes. *Nat. Genet.* **34,** 267–273.

Morley, M., Molony, C. M., Weber, T. M., Devlin, J. L., Ewens, K. G., Spielman, R. S., and Cheung, V. G. (2004). Genetic analysis of genome-wide variation in human gene expression. *Nature* **430,** 743–747.

Moskvina, V., Craddock, N., Holmans, P., Owen, M. J., and O'Donovan, M. C. (2006). *Hum. Hered.* **61,** 55–64.

Moskvina, V., Craddock, N., Holmans, P., Nikolov, I., Pahwa, J. S., Green, E., Wellcome Trust Case Control Consortium Owen, M. J., and O'Donovan, M. C. (2009). Gene-wide analyses of genome-wide association data sets: Evidence for multiple common risk alleles for schizophrenia and bipolar disorder and for overlap in genetic risk. *Mol. Psychiatry* **14,** 252–260.

Myers, S., Bottolo, L., Freeman, C., McVean, G., and Donnelly, P. (2005). A fine-scale map of recombination rates and hotspots across the human genome. *Science* **310,** 321–324.

Newton, M. A., Quintana, F. A., Den Boon, J. A., Sengupta, S., and Ahlquist, P. (2007). Random-set methods identify distinct aspects of the enrichment signal in gene-set analysis. *Ann. Appl. Stat.* **1,** 85–106.

Nicolae, D. L., Gamazon, E., Zhang, W., Duan, S., Dolan, M. E., and Cox, N. J. (2010). Trait-associated SNPs are more likely to be eQTLs: Annotation to enhance discovery from GWAS. *PLoS Genet.* **6,** e1000888.

O'Dushlaine, C., Kenny, E., Heron, E., Donohoe, G., Gill, M., Morris, D., and The International Schizophrenia Consortium Corvin, A. (2010). Molecular pathways involved in neuronal cell adhesion and membrane scaffolding contribute to schizophrenia and bipolar disorder susceptibility. *Mol. Psychiatry* (2010 Feb 16. [Epub ahead of print]).

Oldham, M. C., Konopka, G., Iwamoto, K., Langfelder, P., Kato, T., Horvath, S., and Geschwind, D. H. (2008). Functional organization of the transcriptome in human brain. *Nat. Neurosci.* **11,** 1271–1282.

Pearson, T., Wattis, J. A., O'Malley, B., Pickersgill, L., Blackburn, H., Jackson, K. G., and Byrne, H. M. (2009). Mathematical modelling of competitive LDL/VLDL binding and uptake by hepatocytes. *J. Math. Biol.* **58,** 845–880.

Peters, L. L., Robledo, R. F., Bult, C. J., Churchill, G. A., Paigen, B. J., and Svenson, K. L. (2007). The mouse as a model for human biology: A resource guide for complex trait analysis. *Nat. Rev. Genet.* **8,** 58–69.

Price, A. L., Patterson, N. J., Plenge, R. M., Weinblatt, M. E., Shadick, N. A., and Reich, D. (2006). Principal components analysis corrects for stratification in genome-wide association studies. *Nat. Genet.* **38,** 904–909.

Psychiatric GWAS Consortium Steering Committee (2009). A framework for interpreting genome-wide association studies of psychiatric disorders. *Mol. Psychiatry* **14,** 10–17.

Purcell, S., Neale, B., Todd-Brown, K., Thomas, L., Ferreira, M. A., Bender, D., Maller, J., Sklar, P., de Bakker, P. I., Daly, M. J., *et al.* (2007). PLINK: A tool set for whole-genome association and population-based linkage analyses. *Am. J. Hum. Genet.* **81,** 559–575.

Raychaudhuri, S., Plenge, R. M., Rossin, E. J., Ng, A. C., International Schizophrenia Consortium Purcell, S. M., Sklar, P., Scolnick, E. M., Xavier, R. J., Altshuler, D., *et al.* (2009). Identifying relationships among genomic disease regions: Predicting genes at pathogenic SNP associations and rare deletions. *PLoS Genet.* **5,** e1000534.

Rhee, S. Y., Wood, V., Dolinski, K., and Draghici, S. (2008). Use and misuse of the gene ontology annotations. *Nat. Rev. Genet.* **9,** 509–515.

Richards, A. L., Holmans, P., O'Donovan, M. C., Owen, M. J., and Jones, L. (2008). A comparison of four clustering methods for brain expression microarray data. *BMC Bioinform.* **9,** 490.

Ritchie, M. D., Hahn, L. W., Roodi, N., Bailey, L. R., Dupont, W. D., Parl, F. F., and Moore, J. H. (2001). Multifactor-dimensionality reduction reveals high-order interactions among estrogen-metabolism genes in sporadic breast cancer. *Am. J. Hum. Genet.* **69,** 138–147.

Rual, J. F., Venkatesan, K., Hao, T., Hirozane-Kishikawa, T., Dricot, A., Li, N., Berriz, G. F., Gibbons, F. D., Dreze, M., Ayivi-Guedehoussou, N., *et al.* (2005). Towards a proteome-scale map of the human protein-protein interaction network. *Nature* **437,** 1173–1178.

Schadt, E. E. (2009). Molecular networks as sensors and drivers of common human diseases. *Nature* **461,** 218–223 (Review).

Schadt, E. E., Monks, S. A., Drake, T. A., Lusis, A. J., Che, N., Colinayo, V., Ruff, T. G., Milligan, S. B., Lamb, J. R., Cavet, G., *et al.* (2003). Genetics of gene expression surveyed in maize, mouse and man. *Nature* **422,** 297–302.

Seaman, S. R., and Müller-Myhsok, B. (2005). Rapid simulation of P values for product methods and multiple-testing adjustment in association studies. *Am. J. Hum. Genet.* **76,** 399–408.

Segal, E., Friedman, N., Koller, D., and Regev, A. (2004). A module map showing conditional activity of expression modules in cancer. *Nat. Genet.* **36,** 1090–1098.

Shannon, C. E. (1948). A mathematical theory of communication. *Bell Syst. Tech. J.* **27,** 623–656.

Sherman, B. T., Huang da, W., Tan, Q., Guo, Y., Bour, S., Liu, D., Stephens, R., Baseler, M. W., Lane, H. C., and Lempicki, R. A. (2007). DAVID Knowledgebase: A gene-centered database integrating heterogeneous gene annotation resources to facilitate high-throughput gene functional analysis. *BMC Bioinform.* **8,** 426.

Simes, R. J. (1986). An improved Bonferroni procedure for multiple tests of significance. *Biometrika* **73,** 751–754.

Skol, A. D., Scott, L. J., Abecasis, G. R., and Boehnke, M. (2006). Joint analysis is more efficient than replication-based analysis for two-stage genome-wide association studies. *Nat. Genet.* **38,** 209–213.

Stein, L. D. (2003). Integrating biological databases. *Nat. Rev. Genet.* **4,** 337–345.

Subramanian, A., Tamayo, P., Mootha, V. K., Mukherjee, S., Ebert, B. L., Gillette, M. A., Paulovich, A., Pomeroy, S. L., Golub, T. R., Lander, E. S., and Mesirov, J. P. (2005). Gene set enrichment analysis: A knowledge-based approach for interpreting genome-wide expression profiles. *Proc. Natl. Acad. Sci. USA* **102,** 15545–15550.

Thomas, P. D., Campbell, M. J., Kejariwal, A., Mi, H., Karlak, B., Daverman, R., Diemer, K., Muruganujan, A., Narechania, A. (2003). PANTHER: a library of protein families and subfamilies indexed by function. *Genome. Res.* **13**, 2129–2141

Tian, L., Greenberg, S. A., Kong, S. W., Altschuler, J., Kohane, I. S., and Park, P. J. (2005). Discovering statistically significant pathways in expression profiling studies. *Proc. Natl. Acad. Sci. USA* **102,** 13544–13549.

Torkamani, A., Topol, E. J., Schork, N. J. (2008). Pathway analysis of seven common diseases assessed by genome-wide association. *Genomics* **92**, 265–272

Tsui, I. F., Chari, R., Lam, W. L., and Buys, T. P. (2007). Public databases and software for the pathway analysis of cancer genomes. *Cancer Inform.* **3,** 379–397.

Veyrieras, J. B., Kudaravalli, S., Kim, S. Y., Dermitzakis, E. T., Gilad, Y., Stephens, M., and Pritchard, J. K. (2008). High-resolution mapping of expression-QTLs yields insight into human gene regulation. *PLoS Genet.* **4**(10), e1000214.

Wang, K., Li, M., and Bucan, M. (2007). Pathway-based approaches for analysis of genomewide association studies. *Am. J. Hum. Genet.* **81,** 1278–1283.

Wang, K., Zhang, H., Kugathasan, S., Annese, V., Bradfield, J. P., Russell, R. K., Sleiman, P. M., Imielinski, M., Glessner, J., Hou, C., *et al.* (2009). Diverse genome-wide association studies associate the IL12/IL23 pathway with Crohn disease. *Am. J. Hum. Genet.* **84,** 399–405.

Webber, C., Hehir-Kwa, J. Y., Nguyen, D. Q., de Vries, B. B., Veltman, J. A., and Ponting, C. P. (2009). Forging links between human mental retardation-associated CNVs and mouse gene knockout models. *PLoS Genet.* **5,** e1000531.

Weiss, K. M., and Clark, A. G. (2002). Linkage disequilibrium and the mapping of complex human traits. *Trends Genet.* **18,** 19–24.

Wellcome Trust Case Control Consortium (2007). Genome-wide association study of 14,000 cases of seven common diseases and 3000 shared controls. *Nature* **447,** 661–678.

Xie, X., Lu, J., Kulbokas, E. J., Golub, T. R., Mootha, V., Lindblad-Toh, K., Lander, E. S., and Kellis, M. (2005). Systematic discovery of regulatory motifs in human promoters and 3′ UTRs by comparison of several mammals. *Nature* **434,** 338–345.

Yu, K., Li, Q., Bergen, A. W., Pfeiffer, R. M., Rosenberg, P. S., Caporaso, N., Kraft, P., and Chatterjee, N. (2009). Pathway analysis by adaptive combination of P-values. *Genet. Epidemiol.* **33,** 700–709.

Zaykin, D. V., Westfall, P. H., Young, S. S., Karnoub, M. A., Wagner, M. J., and Ehm, M. G. (2002). Testing association of statistically inferred haplotypes with discrete and continuous traits in samples of unrelated individuals. *Hum. Hered.* **53**(2), 79–91.

Zhang, B., and Horvath, S. (2005). A general framework for weighted gene co-expression network analysis. *Stat. Appl. Genet. Mol. Biol.* **4(Article 17).**

Zhong, H., Yang, X., Kaplan, L. M., Molony, C., and Schadt, E. E. (2010). Integrating pathway analysis and genetics of gene expression for genome-wide association studies. *Am. J. Hum. Genet.* **86,** 581–591.

Zhu, J., Zhang, B., Smith, E. N., Drees, B., Brem, R. B., Kruglyak, L., Bumgarner, R. E., and Schadt, E. E. (2008). Integrating large-scale functional genomic data to dissect the complexity of yeast regulatory networks. *Nat. Genet.* **40,** 854–861.

8

Providing Context and Interpretability to Genetic Association Analysis Results Using the KGraph

Reagan J. Kelly,* Jennifer A. Smith,† and Sharon L. R. Kardia†

*Division of Bioinformatics, ICF International at National Center for Toxicological Research, U.S. Food and Drug Administration, Jefferson, Arkansas, USA

†Department of Epidemiology, School of Public Health, University of Michigan, Ann Arbor, Michigan, USA

ABSTRACT

The KGraph is a data visualization system that has been developed to display the complex relationships between the univariate and bivariate associations among an outcome of interest, a set of covariates, and a set of genetic variations such as single-nucleotide polymorphisms (SNPs). It allows for easy simultaneous viewing and interpretation of genetic associations, correlations among covariates and SNPs, and information about the replication and cross-validation of these associations. The KGraph allows the user to more easily investigate

Advances in Genetics, Vol. 72 0065-2660/10 $35.00

DOI: 10.1016/S0065-2660(10)72008-3

multicollinearity and confounding through visualization of the multidimensional correlation structure underlying genetic associations. It emphasizes gene–environment interactions, gene–gene interactions, and correlations, all important components of the complex genetic architecture of most human traits. The KGraph was designed for use in gene-centric studies, but can be integrated into association analysis workflows as well. The software is available at http://www.epidkardia.sph.umich.edu/software/kgrapher

I. INTRODUCTION

Advances in genetic technologies now allow the assessment of the association between millions of single-nucleotide polymorphisms (SNPs) and complex phenotypes on a routine basis (Ding and Jin, 2009). Genome-wide association studies of complex diseases such as hypertension, diabetes, and heart disease have demonstrated the replicable, significant effects of dozens of genetic polymorphisms. However, these single-gene and single-SNP associations alone do not offer insight into the complex interplay among genes or between genes and the environmental factors that influence phenotypic effects in individuals and populations (Cantor *et al.*, 2010).

Developing methods that aid in interpreting genetic association results within the context of the broader genetic and environmental background is a key problem that must be addressed in the effort to transform the information contained in genetic association analysis results into knowledge that can be applied to understanding a disease's etiology scientifically or to diagnosing or treating a disease clinically (Sing *et al.*, 2003). The biological and metabolic pathways that lead to disease are complex and individual genetic variations, independent of all other contexts, contribute only a small amount of risk for the development or progression of most diseases. Information presented solely at the single-SNP level typically offers only a very limited understanding of the role of a SNP in a disease process (Cantor *et al.*, 2010).

A full exploration of the genetic and environmental context of an individual SNP might place it at the center of a complex network of up-stream factors that influence the SNP's effect and downstream targets influenced by the SNP. As a starting point, therefore, it is useful to shift the focus away from individual SNPs and toward what can be called the "genetic architecture" of the disease phenotypes being explored which provides a more complete picture of the interplay between genetic and nongenetic factors (Sing *et al.*, 1996).

A genetic architecture approach requires exploration of not only the associations of a set of SNPs with the disease phenotype (the outcome of interest), but also their associations with any covariates (i.e., gene–environment correlation), interactions between the covariates and the SNPs (i.e., gene–

environment interactions) that influence the outcome, any pairwise interactions among SNPs (epistasis or gene–gene interactions), and the correlation structure between SNPs (linkage disequilibrium (LD)) and between covariates (environmental correlations). To provide this broader view of the genetic and environmental context to aid in interpreting results from association studies, we describe the "KGraph," a method of displaying association, interaction, and correlation information about a set of SNPs and covariates being investigated with regard to a particular outcome of interest originally presented by Kelly *et al.* (2007).

II. THE ROLE OF CORRELATION STRUCTURE IN INTERPRETING RESULTS

Measures of direct association provide insight into the relationship of predictors (SNPs or covariates) and outcomes, but these associations do not necessarily correspond to direct causal relationships. Often there are important confounders that associations and a deeper investigation into the correlation structure underlying the predictors can clarify the relationships among the predictors and suggest useful avenues for further study.

It is well known that even very strong association between an individual SNP and an outcome does not necessarily identify that particular SNP as causal. It is just as likely that a SNP in LD with the investigated SNP may have a causal relationship with the outcome. Estimating pairwise LD between SNPs can help refine this knowledge. If a group of SNPs that are associated with the outcome have high pairwise LD among themselves, it suggests that a haplotype block containing these SNPs may harbor a SNP with a true causal effect on the outcome. Using an external resource like the HapMap, the full set of SNPs that make up that haplotype block can be identified and used as a starting point for molecular research to fully quantify the relationship between this set of SNPs and the outcome (International HapMap Consortium *et al.*, 2007).

As with SNPs, the correlation between pairs of covariates can offer insight into the relationships that the covariates have with the outcome. For example, if two covariates with moderate correlation have differing associations with the outcome (i.e., one is strongly associated while the other has no association), this suggests that the information contained in the associated covariate that is not contained in the unassociated covariate is driving the association. By considering the biological relationship of the two covariates, a testable hypothesis about the effect of the associated covariate on the outcome can be generated.

Correlation structure extends beyond simple pairwise correlation between predictors of the same type. It is also important to consider the relationships between SNPs and covariates in order to more fully characterize the relationship between SNPs and the outcome. If a SNP that is highly associated with the outcome is also highly associated with a covariate, and if the covariate is in turn highly associated with the outcome, this suggests either that the SNP association is simply marking the covariate's association, or potentially that the SNP is part of a causal chain that affects the covariate which in turn affects the outcome.

III. IMPROVING INTERPRETABILITY BY INCLUDING CONTEXT-DEPENDENT EFFECTS

Considering only single-SNP associations misses a potentially large and valuable set of context-dependent results (Cordell, 2009; Hebebrand and Hinney, 2009; Snieder *et al.*, 2008). These SNPs only have a significant association when considered as part of a SNP–covariate or SNP–SNP interaction. Increasing evidence suggests that these effects are the norm rather than the exception (Moore, 2003).

It is important, however, to draw a distinction between "statistical" interactions and "biological," or mechanistic, interactions. Even a highly significant statistical interaction does not guarantee there is an underlying biological mechanism that includes the two members of the interaction. The significance of the interaction could be due simply to a correlation whose true cause is still obscure. Functional annotation of SNPs and covariates, however, can provide a starting point for biological interpretation. For example, if interacting SNPs are found in two genes known to physically interact, further exploration might show that the polymorphisms being present together prevent their binding. The statistical results, however, are a necessary starting point and potential way to form testable biological hypotheses (Moore, 2003).

IV. OVERVIEW OF THE KGRAPH'S STRUCTURE

The KGraph was designed as a tool to aid in visual exploration of SNP association results in the context of relevant genetic and environmental information. An example of a KGraph is shown in Fig. 8.1. The KGraph has eight graphical regions, each of which displays the results from a single type of statistical analysis. These regions are arranged into two major sections to show (1) the interrelationships among genetic and covariate associations with the outcome, and (2) the underlying correlation among these genetic factors and covariates. The inner section displays the underlying correlation structure in the form of SNP–SNP LD, SNP–covariate association, and covariate–covariate correlation. The outer section displays associations with the outcome of interest (i.e., single-covariate association, single-SNP association, covariate–covariate interaction, SNP–covariate interaction, and SNP–SNP interaction).

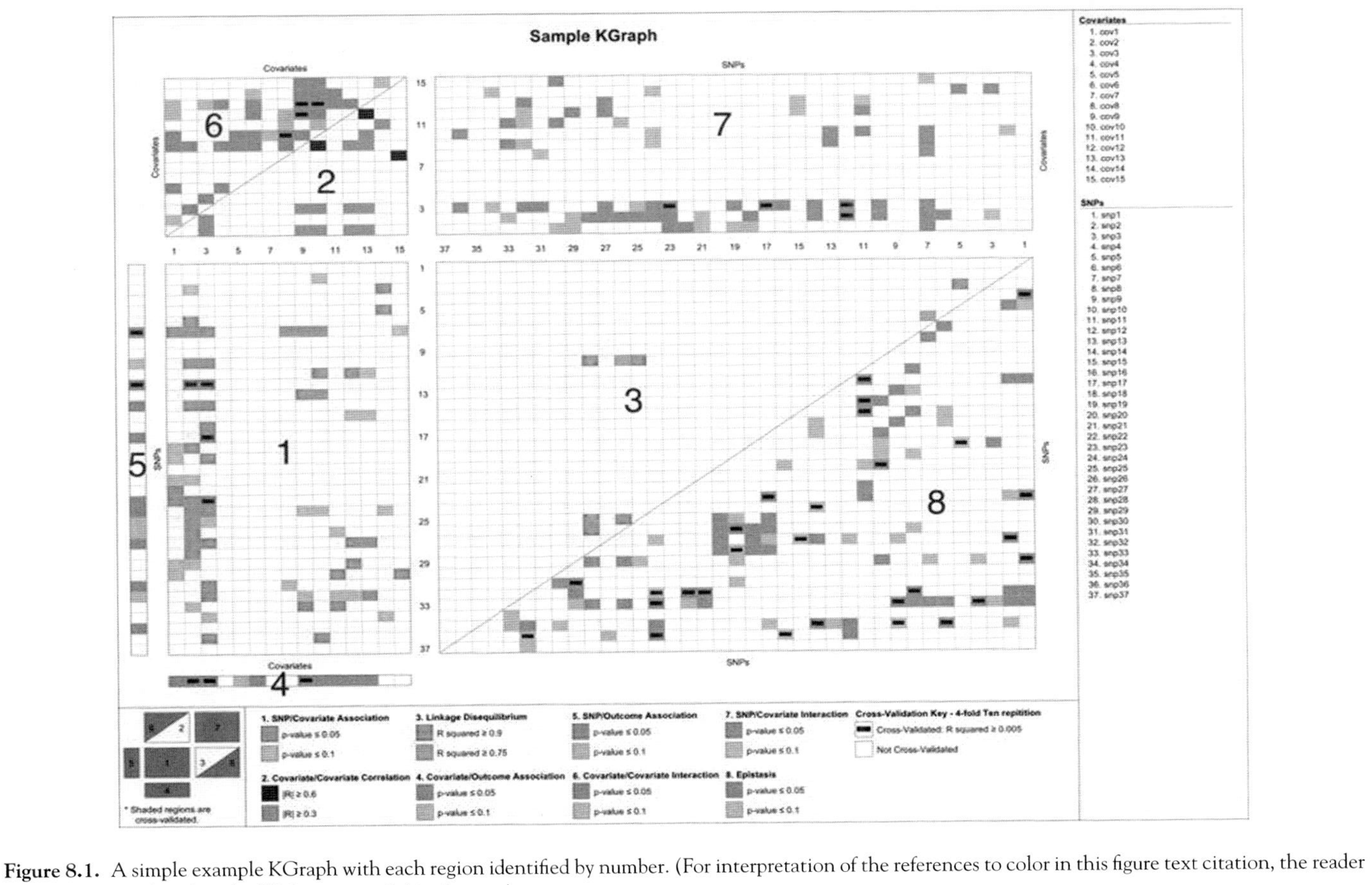

Figure 8.1. A simple example KGraph with each region identified by number. (For interpretation of the references to color in this figure text citation, the reader is referred to the Web version of this chapter.)

The correlation structure displayed by the inner section provides the background that allows the association results to be interpreted in terms of the relationships between and among the SNPs and covariates. The first region, labeled 1 in Fig. 8.1, displays the association between the SNPs and covariates. This is typically assessed by an analysis of variance (ANOVA) for continuously distributed covariates and chi-square tests for categorical covariates; however, users may display the results from any association measurement of their choice. Cells representing significant results are colored light green while those representing highly significant results are dark green. Region 2 displays covariate–covariate correlations with moderate correlations colored light gray and strong correlations colored dark gray. Correlation is commonly measured using a correlation coefficient or a chi-square test *p*-value; however, users can assess correlation with their preferred test and define the correlation levels that are considered “strong” and “moderate.” LD among SNPs is displayed in Region 3, with strong disequilibrium shaded as dark red and moderate disequilibrium shaded light red. LD can be assessed using any standard measure, such as D, D', or R^2, and as with covariate–covariate correlation users can define the level of LD to be considered “strong” and “moderate.”

The outer section displays the association between the outcome of interest and the covariates and SNPs being examined. For consistency, all of the regions in the outer section cells representing significant associations are colored light blue and cells representing highly significant associations are dark blue. Region 4 displays the association between the covariates and the outcome of interest as measured by an appropriate modeling technique (e.g., logistic regression or linear regression) depending on the type of outcome. Region 5 displays the association between the SNPs and the outcome. The remaining three sections display the results from testing for first-order interactions between two covariates (Region 6), between pairs of SNPs and covariates (Region 7), or between two SNPs (Region 8). The significance of associations is typically expressed as a *p*-value; however, as with the correlation section of the KGraph, users can substitute different measures, such as FDR *q*-value, to their preference.

While association analyses are an important way to identify potentially interesting SNPs, the growing number of SNPs being investigated increases the number of false-positive findings. To assist in differentiating false positives from true positives, it has become common practice to estimate a significant SNP’s ability to make a prediction in an independent set of data using cross-validation or determine if the same association is replicated in a different dataset (Manly, 2005; Molinaro *et al.*, 2005). To account for this, the KGraph can represent both cross-validation (indicated by a small horizontal bar within the cell) and/or replication results (displayed by dividing a cell with a diagonal line, with each half-cell representing one sample). Again, the KGraph system is highly flexible, and while the horizontal bar marking cells was originally intended to indicate cross-validation, users can utilize it to represent any dichotomous criteria they choose.

V. CREATING AND EXPLORING KGRAPHS

In order to easily construct KGraphs from association analysis results, we have developed a creation and display utility named KGrapher. The KGrapher utility provides a "wizard interface" that prompts the user for the files and criteria needed to plot each region. The input files are the results from statistical analyses. The criteria include the names of the SNPs and covariates, the significance measurements desired by the user (e.g., ANOVA *p*-value, Pearson's correlation coefficient, etc.), the alpha levels used to designate significant and highly significant results, and the cross-validation measurement and cutoff values.

In addition to creating KGraphs, KGrapher can be used as a full-featured KGraph viewer that allows interactive exploration. KGrapher provides an innovative overlay and a tool-tip system to allow users to easily identify specific results, access information about significance tests and cross-validation, and examine patterns in tests involving the same predictors. KGrapher also allows the user to generate a number of KGraphs simultaneously using a batch creation mode.

KGrapher accepts both tab delimited (.txt) and comma separated (.csv) files for data input. It uses a dedicated file format (.kgr) that contains both the input data and the configuration settings for a given KGraph. Users are able to easily save KGraphs, print them, and share them with other researchers. Additionally, KGrapher allows users to export KGraphs as high-resolution JPEG (.jpg) files for publication or display on the Internet.

Although the concept of using different color intensities to represent differing levels of significance was inspired by the heatmap displays common in gene expression analysis, the choice was made to restrict the number of colors to two. A complete spectrum of possible colors would be difficult to interpret, and so in the interest of simplicity only two color levels are available. Precise comparison between two results can be performed by using the hover and tool-tip functionalities built into the KGrapher utility.

VI. EXPLORING GENETIC ARCHITECTURE IN REAL-WORLD DATA USING A KGRAPH

A real-world example of a KGraph being used in a candidate gene study comes from Smith *et al.* (2009). In order to identify SNPs that affect hypertensive damage to white matter in the brain, the Genetic Epidemiology Network of Arteriopathy (GENOA), a community-based study of hypertensive sibships that aims to identify genes influencing blood pressure, undertook the Genetics of Microangiopathic Brain Injury (GMBI) substudy. Seven hundred and seventy seven participants underwent magnetic resonance imaging (MRI) of the brain in order to quantify leukoaraiosis volume, or the volume of ischemic damage to the subcortical white matter of the brain. Participants were also genotyped for 1649 SNPs from genes known or hypothesized to be involved in arteriosclerosis and related pathways.

The statistical analysis performed included modeling SNP main effects, epistatic interactions, and gene–environment interactions between these SNPs and covariates (including conventional and novel risk factors for arteriosclerosis) for association with leukoaraiosis volume. Three methods were used to reduce the chance of false-positive associations: (1) false discovery rate (FDR) adjustment for multiple testing, (2) an internal replication design, and (3) a 10-iteration fourfold cross-validation scheme.

Figure 8.2 shows the KGraph created from these analyses. A key to the eight regions of the KGraph, giving the criteria used for shading and a description of each region, is located in the lower left corner of Fig. 8.2. Included on the KGraph are all of the covariates that were investigated in the study, SNPs that were involved in a single-SNP or SNP–covariate association that passed all three statistical filters (FDR q-value <0.3, association p-value <0.1 in both replication datasets, and cross-validation $R^2>0.5\%$), and SNPs that were involved in at least one of the 20 most highly predictive SNP–SNP interactions that passed all three filters. All associations involving these SNPs and covariates are presented on the KGraph, and those that passed all three filters are indicated by a horizontal black bar.

Region 1 in Fig. 8.2, shown in green, displays the association between the SNPs and covariates. The majority of SNP–covariate associations were accounted for by three SNPs in the factor VIII (*F8*) gene that were associated with log serum creatinine, height, HDL cholesterol, waist-to-hip ratio, and weight. Region 2, shown in gray, illustrates the correlations between the covariates. The majority of the risk factors are significantly associated with one another (p-value <0.05). Region 3, in red, shows the observed LD, estimated in a sample of 357 unrelated individuals from the study sample. As expected, significant LD ($r^2>0.5$) occurs only between SNPs that are within the same gene.

The remaining regions are colored blue, indicating that they represent associations with the outcome of interest, leukoaraiosis. Region 4, which displays the univariate association between the covariates and leukoaraiosis, shows that only age had an association that met all the three criteria. Region 5, which illustrates univariate associations between the SNPs and leukoaraiosis, shows that four SNPs have significant, replicated, and cross-validated associations (*F3_rs3917643*, *CAPN10_rs7571442*, *MMP2_rs9928731*, *KITLG_rs995029*). These single-SNP and single-covariate associations do not provide much insight into the genetic architecture of leukoaraiosis on their own. However, adding interactions provides a fuller picture.

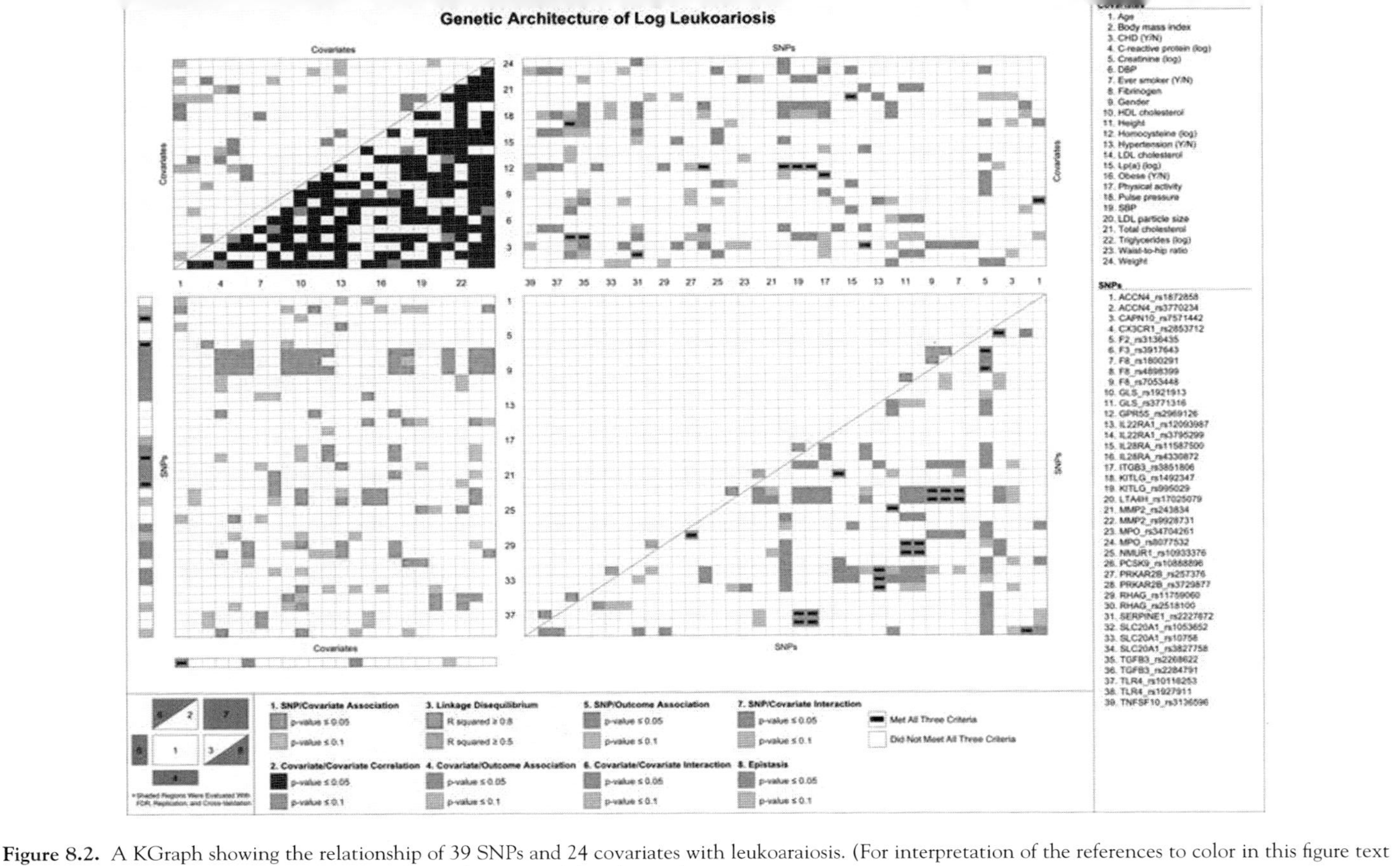

Figure 8.2. A KGraph showing the relationship of 39 SNPs and 24 covariates with leukoaraiosis. (For interpretation of the references to color in this figure text citation, the reader is referred to the Web version of this chapter.)

Region 6 displays the covariate–covariate interactions that are significantly associated with leukoaraiosis, but no interactions of this type passed all three filters. Region 7 displays the interactions between the SNPs and covariates that were associated with leukoaraiosis. Overall, 12 interactions were detected that replicated and cross-validated, though two pairs of SNP–covariate interactions appear to be marking the same association due to strong LD between the involved SNPs. Two SNPs in *KITLG* (SNPs 18 and 19 in Fig. 8.2) have a significant, cross-validated interaction with homocysteine; however, 3 shows that SNPs 18 and 19 are in strong LD. Likewise, two SNPs in *TGFB3* (SNPs 35 and 36 in Fig. 8.2) have a significant, cross-validated interaction with C-reactive protein level, but 3 shows them to be in strong LD.

Region 8 displays the epistatic (SNP–SNP) interactions significantly associated with leukoaraiosis. One hundred seventy three replicated and cross-validated interactions were identified. The most predictive interactions included those between SNPs in *RHAG* and *GLS*, *F8* and *MPO*, *SLC20A1* and *IL22RA*, *KITLG* and *TLR4*, *NMUR1* and *GPR55*, *ACCN4* and *TNFSF10*, and *CX3CR1* and *F2*. Interactions between two genes that appear more than once in the SNP–SNP results are almost entirely due to strong LD between involved SNPs. This can be seen most clearly in three SNPs from *F8* (SNPs 7, 8, and 9 in Fig. 8.2), which have high pairwise LD, and have significant, cross-validated interactions with two SNPs from *MPO* (SNPs 21 and 22 in Fig. 8.2) that also have moderate pairwise LD.

SNPs in *CAPN10* and *F3* show significant, cross-validated association with leukoaraiosis univariately, but have no significant, cross-validated interactions with covariates or other SNPs. This suggests that these SNPs possibly affect leukoaraiosis directly rather than through a context-dependent mechanism. Moreover, these SNPs have no significant, cross-validated association with the investigated covariates, suggesting that their effect is likely not through a pathway involving these covariates. Two SNPs in *KITLG* show significant, cross-validated interactions with two SNPs from *TLR4*. Although this is statistical rather than biological epistasis, further investigation into the mechanisms of these genes may provide insight into their relationship with leukoaraiosis.

To demonstrate the increased ability to predict leukoaraiosis volume resulting from this broader analysis including both SNP–covariate and SNP–SNP interactions, two predictive models were constructed. The first model, which included only the four SNPs which were identified as significant after filtering for false-positive results, had a cross-validation R^2, a measure of a model's ability to predict the outcome, of 3.72%. A second model that contained the four single-SNP associations, the four most highly significant SNP–covariate interactions, and the four most significant SNP–SNP interactions had a cross-validation R^2 of 11.6%. These results indicate that the genetic architecture of complex traits such as leukoaraiosis is comprised of SNP and covariate main effects, gene–gene interactions, and gene–environment interactions from a variety of biological pathways.

Although genome-wide studies examine a much larger number of SNPs than would be practical to display fully, KGraphs still offer an attractive way to contextualize results. The user could identify an interesting subset of results, for instance a set of SNPs that has been replicated in several genome-wide studies of the same trait. After identifying other polymorphisms within 5 kb of the replicated set of SNPs, the complete set of SNP–SNP and SNP–covariate analyses could be performed and a KGraph created from the results. The KGraph's ability to display results from two datasets simultaneously by dividing each cell diagonally would also provide a way to visually identify results that replicate in two different samples. In order to more fully examine pleiotropic effects, results from two different traits in the same dataset can also be compared by dividing the cells diagonally.

VII. ENHANCEMENTS AND FUTURE DIRECTIONS FOR THE KGRAPH

Although the KGraph and KGrapher utility are stable technologies that greatly assist the interpretation of genetic association studies, there are several useful features that could be added to enhance their usefulness. More direct integration into analysis workflows, advanced selection techniques that simplify the display of large numbers of SNPs, and data interpretation tools that combine public data with association results would improve the utility and appeal of KGraphs.

Significant effort was put into making KGraphs easy to create for scientists and researchers. However, it still requires additional steps above and beyond the typical analysis. Because the KGrapher utility was created with a batch-processing interface that can operate noninteractively, it would be possible to directly connect KGraph creation to analysis. Assuming that the necessary analysis files have been created using a statistical language such as *R*, it would be straightforward to write scripts that create the appropriate configuration files and then call the KGrapher utility to create and export KGraphs. It would be more complex, but still practical, to integrate KGraph creation into analysis workflows. For example, an *R* script could be created that selected SNPs according to user-specified criteria, performed the necessary additional analyses, created the appropriate files, and automatically generated the KGraph.

For very high dimensional genetic association studies, including the results from genome-wide associations, it is possible that the correlation structure that forms the core of the KGraph could be leveraged to reduce the number of SNPs being displayed to a more manageable number. Rather than forcing the user to winnow the set of SNPs, it is possible to add powerful filtering criteria into the KGrapher utility that would allow users to automate this process by creating rules (e.g., "For a set of SNPs with a minimum pairwise LD r^2 of 0.75 display only the SNP with the largest cross-validated R^2" or "Show only SNPs with two or more significant, cross-validated SNP–SNP interactions"). For instance, automated filtering criteria would allow the user to more fully examine the gene–environment interactions, epistasis, gene–environment correlations, and LD of all variations within 5 kb of the top, replicated genomic regions that have been found in GWAS consortia.

The KGrapher utility could also add another layer of context to the KGraph. By connecting with publicly available pathway databases such as KEGG and other data sources such as dbSNP, the polymorphisms identified as significant and cross-validated could be seen in the context of the biological pathway they affect. For example, KGrapher could search KEGG and identify the pathways that each SNP with one or more significant, cross-validated result belongs to, and then display that list ordered by the number of SNPs found in each pathway. KGrapher could also visually indicate which SNPs were implicated in a particular selected pathway. This would provide needed biological context to results and help form testable hypotheses about the specific effect of a SNP on a gene.

VIII. CONCLUSION

Displaying genetic association results graphically in a unified, coherent way allows a researcher to identify and more fully characterize SNPs that act in a context-dependent manner. Although designed with candidate gene studies in mind, intelligent choices about the SNPs to include give it utility in genome-wide association studies as well.

The additional context provided by the KGraph can enable researchers to form testable hypotheses about the role of SNPs and to determine when a given SNP should be considered in a predictive model. Additional enhancements to the KGraph format and the KGrapher utility deserve focus and could make the tool more powerful and simpler to integrate into an analysis. The most recent version can be obtained at http://www.epidkardia.sph.umich.edu/software/kgrapher.

References

Cantor, R. M., Lange, K., and Sinsheimer, J. S. (2010). Prioritizing GWAS results: A review of statistical methods and recommendations for their application. *Am. J. Hum. Genet.* **86,** 6–22.

Cordell, H. J. (2009). Detecting gene–gene interactions that underlie human diseases. *Nat. Rev. Genet.* **10,** 392–404.

Ding, C., and Jin, S. (2009). High-throughput methods for SNP genotyping. *Methods Mol. Biol.* **578,** 245–254.

Hebebrand, J., and Hinney, A. (2009). Environmental and genetic risk factors in obesity. *Child Adolesc. Psychiatr. Clin. N. Am.* **18,** 83–94.

International HapMap Consortium Frazer, K. A., Ballinger, D. G., Cox, D. R., Hinds, D. A., Stuve, L. L., Gibbs, R. A., Belmont, J. W., Boudreau, A., Hardenbol, P., *et al.* (2007). A second generation human haplotype map of over 3.1 million SNPs. *Nature* **449,** 851–861.

Kelly, R. J., Jacobsen, D. M., Sun, Y. V., Smith, J. A., and Kardia, S. L. R. (2007). KGraph: A system for visualizing and evaluating complex genetic associations. *Bioinformatics* **23,** 249–251.

Manly, K. F. (2005). Reliability of statistical associations between genes and disease. *Immunogenetics* **57,** 549–558.

Molinaro, A. M., Simon, R., and Pfeiffer, R. M. (2005). Prediction error estimation: A comparison of resampling methods. *Bioinformatics* **21,** 3301–3307.

Moore, J. H. (2003). The ubiquitous nature of epistasis in determining susceptibility to common human diseases. *Hum. Hered.* **56,** 73–82.

Sing, C. F., Haviland, M. B., and Reilly, S. L. (1996). Genetic architecture of common multifactorial diseases. *Ciba Found. Symp.* **197,** 211–229, discussion 229–232.

Sing, C. F., Stengard, J. H., and Kardia, S. L. (2003). Genes, environment, and cardiovascular disease. *Arterioscler. Thromb. Vasc. Biol.* **23,** 1190–1196.

Smith, J. A., Turner, S. T., Sun, Y. V., Fornage, M., Kelly, R. J., Mosley, T. H., Jack, C. R., Kullo, I. J., and Kardia, S. L. (2009). Complexity in the genetic architecture of leukoaraiosis in hypertensive sibships from the GENOA study. *BMC. Med. Genomics* **2,** 16

Snieder, H., Wang, X., Lagou, V., Penninx, B. W., Riese, H., and Hartman, C. A. (2008). Role of gene–stress interactions in gene-finding studies. *Novartis Found. Symp.* **293,** 71–82, discussion 83–86, 122–127.

Index

W

X